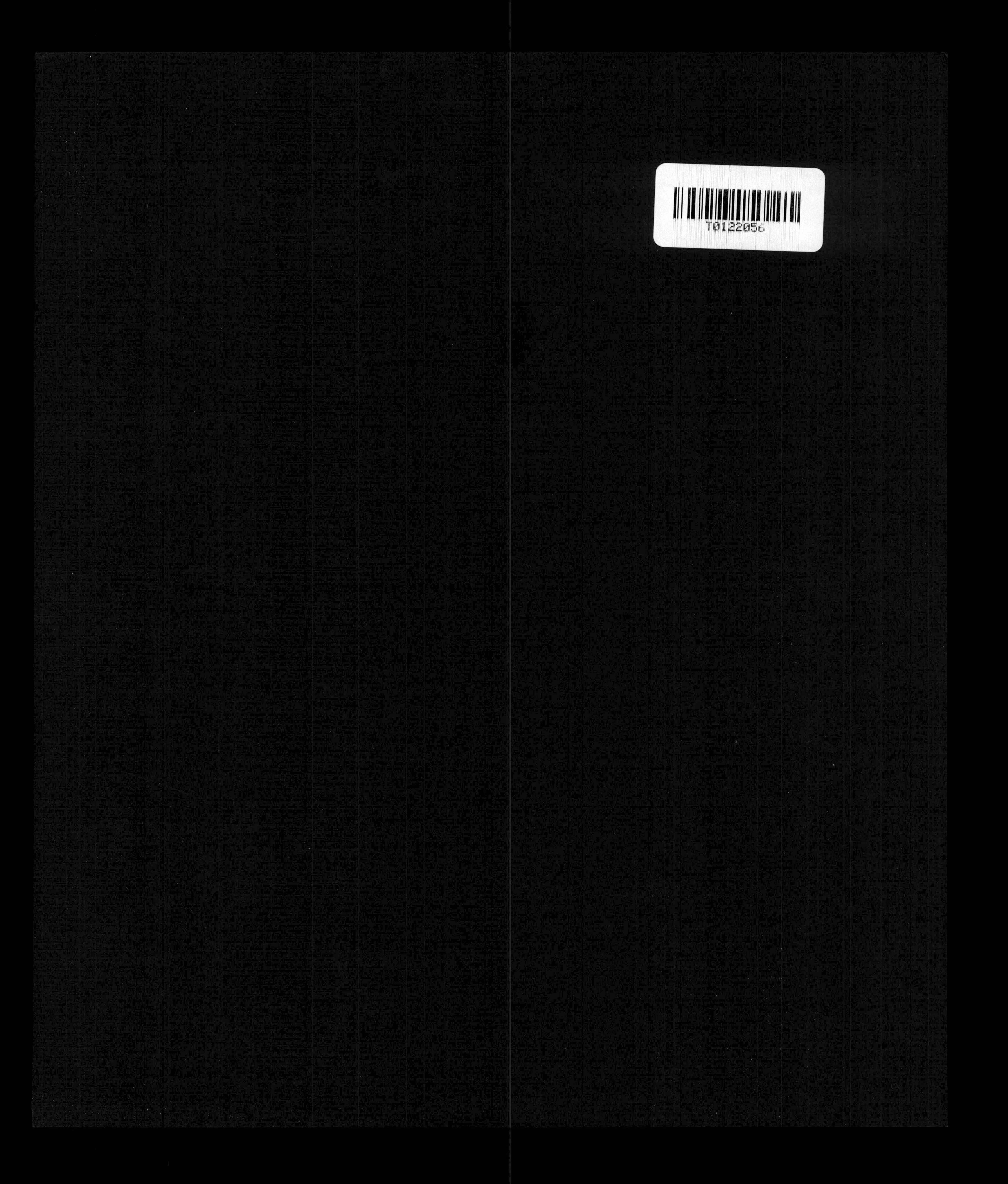
T0122056

Out of this World

Historic Milestones in NASA's Human Space Flight

Discovery

Out of this World

Historic Milestones in NASA's Human Space Flight

Bill Schwartz

ACC ART BOOKS

This book is dedicated to my father;
for always being inspirational and kind-hearted

Introduction

On a mission to uncover the mysteries of our universe that took place after the Big Bang, as well as landing astronauts back on the moon, NASA continues to take large steps forward in the world of space exploration. With the recent successful launch of the James Webb Telescope in December 2021, and the upcoming Artemis SLS rocket missions due to take place sometime in 2022, NASA's perseverance, passion, and extraordinary history of overcoming all odds, is apparent for all to see.

In 1969 over 650 million people around the world simultaneously witnessed man's first footsteps on the moon. The universal feeling of optimism, the shared pride in one man's footsteps, the hope and the astonishment were almost tangible. It was unfathomable. It was a fleeting moment in time when all religious or political conflicts were overshadowed by a sense of pure humanity.

I was nine years old in 1969, and fondly remember watching The Jetsons and Star Trek with my dad on a black-and-white television set in Santa Monica, California. When the moon landing came on our television screen, I was utterly enthralled by the grainy images of NASA astronaut Neil Armstrong setting foot on the moon and uttering, possibly, the most famous phrase in the history of mankind 'That's one small step for man. One giant leap for mankind.' The simplicity of those words belied the magnitude of that momentous event.

I collected every *Life* and *National Geographic* magazine I could lay my hands on, to satisfy my insatiable appetite for the wonders of space exploration. The story told by a single printed image had a far deeper meaning and greater impact on me than watching film footage of the event. The intense expression on the face of a young astronaut waiting for liftoff, the Earth suspended in blackness, the splashdown of a successful mission; all these moments frozen in time enabled me to be there 'in the moment', rather than being just a passive viewer. This book, filled with original NASA high-resolution photographs, paintings and illustrations paired with wonderful quotes, allows the reader to open any page, to pause and to ponder.

The fuzzy broadcast images of every space mission caused me (and much of the world) to gasp in wonderment. I was always amazed when listening to interviews with the astronauts who seemed so young and yet were so sharp and so unaffected by all the attention. CBS's Walter Cronkite, the most trusted newsman in America, was the voice, eyes and ears of every NASA mission. His enthusiasm was infectious – so much so that it felt as if he took me with him, making each space exploration voyage my own personal adventure. When I think of the broadcast coverage of NASA's milestone missions to the moon, the Skylab, the space shuttles, the International Space Station and missions to deep space, it takes me back to moments of intense suspense and the feeling of sheer terror for the astronauts, even as a spectator. Each spacewalk, moon landing and space-station addition writing another chapter in NASA's incredible history.

NASA has now celebrated its diamond anniversary: 60 incredible years that have seen the launch of over 200 manned space-bound flights. This book is a celebration of those magnificent six decades of space exploration. We don't need scientific jargon and technical data to appreciate the magnitude of NASA's missions. Never has the phrase 'a picture paints a thousand words' been more apt than in describing the images that I have selected for this book. It was my great honor to secure permission to use quotes from the likes of astronaut James Lovell, commander of the aborted Apollo 13 mission and Dr. Neil deGrasse Tyson, famed astrophysicist and planetary scientist. I have also included the words and thoughts of students, baristas, theologians, grandparents and the very young, all of which encapsulate the mysteries and the wonders of space exploration.

When I was young, there was no high-speed internet and there were no *Star Wars* movies! Thinking back to that time, it is extraordinarily difficult to envisage how one could even contemplate putting a man on the moon. Driven by the charisma and all-out determination of President Kennedy to win the space race against Russia, and by over 400,000 workers who pooled their resources, a landing on the moon transitioned from being a possibility to becoming a reality.

Striking images of President Kennedy inspecting rockets and giving his famed Rice University speech, alongside images of intrepid astronauts risking their all, reinforces the very real human element of this process amidst all the science and technology. Photographs from all the Apollo missions bring to life the sense of adventure while also emphasizing the vulnerability of the astronauts. High-resolution images, such as the famous Lunar Moon Rover with its duck-tape-repaired fender, and the extremely cramped living quarters of astronauts and fabulous 'splash-downs', will certainly spark the reader's imagination. As space is indeed the final frontier, images of the earth taken from the moon are exceptionally emotive and powerful, clearly demonstrating the vulnerability of our tiny planet Earth in a sea of blackness. They bring a perspective that is even more meaningful now with the very real potential of catastrophic climate change. To view the Earth alone is to see humanity hanging by a thread.

In reliving the excitement of the "space-shuttle era" from 1981 to 2011, with its 135 missions, I felt I was revisiting an old friend. The space shuttle took on a personality like no other space-craft in history. While the tragedies and the loss of life were devastating, the monumental capabilities of the space shuttle were extraordinary. Amongst other things, it was instrumental in the piece-by-piece construction of the International Space Station and the launch of the Hubble Space Telescope and satellites. This 'space delivery vehicle' was instrumental in taking space exploration into a new era. I am pleased to share images of the space shuttle, including a photo of me witnessing the Space Shuttle Endeavour's last journey through the streets of Los Angeles on 12 October 2012, to its final resting place at the Samuel Oschin Space Shuttle Endeavour Pavilion at the California Science Center. It's one thing to see a video of the space shuttle, but seeing it 'in person' was another thing altogether. It was extraordinarily thrilling.

Some may regard the International Space Station as an 'intermediate' space colony until such time as the Moon or Mars is colonized. Since 1998, thousands of experiments have been conducted on the ISS. This mighty platform is 356 feet end-to-end, has eight miles of wiring and 43,000 cubic feet of habitable, pressurized volume (which is about the volume of three average American houses). Usually, there are around seven astronauts on board. The technology created and used by the International Space Station is the foundation for future colonies and commercialization of space.

Woven through this book are a series of portraits which honor the astronauts, politicians, scientists and inventors who are the faces and brain thrust of NASA. It is an administration which is only as good as the men and women on whose shoulders its greatest achievements rest. NASA has been at the forefront of offering opportunities to all, without prejudice. With this in mind, I look forward to seeing the first female astronaut setting foot on the moon! Space exploration, space technology, Earth and space science, and aeronautics research all fall under the umbrella of NASA. For this reason, I have included images of deep space from the Hubble Space Telescope which can identify the light of a firefly at some 7,000 miles. Recent images contain more than 5,000 galaxies, some of them as far as 13.2 billion light years away. The James Webb Space Telescope, which is 100 times more powerful than the Hubble telescope, has the potential to rewrite our textbooks.

It is my wish that, with the jaw-dropping images and inspiring quotes from *Out of this World*, both young and old will have a joyful journey. At the same time, I hope that the journey will spark hope and humanitarianism in all who read this book.

1: JFK and the Space Race

I remember being six or seven years old and the teacher at Canfield Elementary School in Los Angeles, California, would periodically interrupt the lesson and all of a sudden yell "Drop and Cover!" at which point all the students would immediately drop to the floor and cover their heads under the desk in preparation for a potential nuclear bomb from the Soviet Union. I now think about the insanity of the whole thing on so many different levels! However, it is fascinating to think that, since the end of WWII, there existed a very adversarial race to gain strategic advantage in space for national security, and increase the global "super-power" persona of the United Sates and the Soviet Union. The turning point came on May 25th, 1961...

Above: United States President John F. Kennedy outlines his vision for manned exploration of space to a Joint Session of the United States Congress, in Washington DC, on 25 May 1961. Shown in the background are, (left) Vice President Lyndon Johnson, and (right) Speaker of the House.

[Photo and caption credit: MediaPunch Inc / Alamy Stock Photo]

A Grand Vision

There was a sense of anticipation in the United States at the beginning of 1961 when John F. Kennedy was sworn in as the nation's new president and he promised to reach for a "New Frontier." However, the Cold War soon took center stage when news of the Soviet cosmonaut Yuri Gagarin becoming the first human in space stunned Americans on 12 April 1961. Just a few days later, the nation and the new administration were embarrassed once again by the Bay of Pigs fiasco. These events put unquantifiable pressure on the new president. Four years earlier, Americans had been shocked when the Soviets orbited the first satellite and the USA was still lagging behind. Kennedy responded with a proposal that would inspire the nation and the world.

After consulting with Vice President Lyndon B. Johnson, NASA administrator James Webb and other officials, Kennedy sought to challenge the nation's rival to a new race. The president concluded that landing an American on the Moon would be a technological feat that could erase the Soviet Union's head start. It was a new beginning, giving the USA an opportunity to demonstrate the superiority of democracy over communism.

On 5 May 1961, NASA astronaut Alan Shepard flew on a short sub-orbital flight into space. While not orbiting the Earth like Gagarin, the 15-minute trip down the Atlantic Missile Range was a success Kennedy could build upon.

A mere three weeks after Shepard's flight, President Kennedy went before a special joint session of Congress and announced a dramatic and ambitious goal. "I believe this nation should commit itself to achieving the goal, before this decade is out, of landing a man on the Moon and returning him safely to Earth," he said.

The decision involved much deliberation before making it public, due to the enormous human efforts and expenditures it would entail to make what became Project Apollo a reality—and to do so by 1969. Only the construction of the Panama Canal in modern peacetime and the Manhattan Project to develop the atomic bomb in war were comparable in scope.

Many historians now view Kennedy's drive to send men to the Moon as a political decision as much as one of adventure and exploration.

Much of NASA's early human spaceflight efforts were guided by Kennedy's leadership. Projects Mercury in its latter stages, Gemini, and Apollo were all designed to meet his objective. His goal was achieved on 20 July 1969, when Apollo 11 commander Neil Armstrong stepped off the lunar module's ladder and onto the Moon's surface. The world watched in awe.

The momumental JFK speech to the population which ignited the passions, vision, and race to space:

President Pitzer, Mr. Vice President, Governor, Congressman Thomas, Senator Wiley and Congressman Miller, Mr. Webb, Mr. Bell, scientists, distinguished guests, and ladies and gentlemen:

I appreciate your president having made me an honorary visiting professor, and I will assure you that my first lecture will be very brief. I am delighted to be here, and I'm particularly delighted to be here on this occasion.

We meet at a college noted for knowledge, in a city noted for progress, in a State noted for strength, and we stand in need of all three, for we meet in an hour of change and challenge, in a decade of hope and fear, in an age of both knowledge and ignorance. The greater our knowledge increases, the greater our ignorance unfolds.

Despite the striking fact that most of the scientists that the world has ever known are alive and working today, despite the fact that this Nation's own scientific manpower is doubling every 12 years in a rate of growth more than three times that of our population as a whole, despite that, the vast stretches of the unknown and the unanswered and the unfinished still far outstrip our collective comprehension.

No man can fully grasp how far and how fast we have come, but condense, if you will, the 50,000 years of man's recorded history in a time span of but a half-century. Stated in these terms, we know very little about the first 40 years, except at the end of them advanced man had learned to use the skins of animals to cover them. Then about 10 years ago, under this standard, man emerged from his caves to construct other kinds of shelter. Only five years ago man learned to write and use a cart with wheels. Christianity began less than two years ago. The printing press came this year, and then less than two months ago, during this whole 50-year span of human history, the steam engine provided a new source of power.

Newton explored the meaning of gravity. Last month electric lights and telephones and automobiles and airplanes became available. Only last week did we develop penicillin and television and nuclear power, and now if America's new spacecraft succeeds in reaching Venus, we will have literally reached the stars before midnight tonight.

Above: US President John F. Kennedy delivers his famous speech on space

This is a breathtaking pace, and such a pace cannot help but create new ills as it dispels old, new ignorance, new problems, new dangers. Surely the opening vistas of space promise high costs and hardships, as well as high reward.

So, it is not surprising that some would have us stay where we are a little longer to rest, to wait. But this city of Houston, this state of Texas, this country of the United States was not built by those who waited and rested and wished to look behind them. This country was conquered by those who moved forward—and so will space.

William Bradford, speaking in 1630 of the founding of the Plymouth Bay Colony, said that all great and honorable actions are accompanied with great difficulties, and both must be enterprised and overcome with answerable courage.

If this capsule history of our progress teaches us anything, it is that man, in his quest for knowledge and progress, is determined and cannot be deterred. The exploration of space will go ahead, whether we join in it or not, and it is one of the great adventures of all time, and no nation which expects to be the leader of other nations can expect to stay behind in the race for space.

Those who came before us made certain that this country rode the first waves of the industrial revolutions, the first waves of modern invention, and the first wave of nuclear power, and this generation does not intend to founder in the backwash of the coming age of space. We mean to be a part of it—we mean to lead it. For the eyes of the world now look into space, to the moon and to the planets beyond, and we have vowed that we shall not see it governed by a hostile flag of conquest, but by a banner of freedom and peace. We have vowed that we shall not see space filled with weapons of mass destruction, but with instruments of knowledge and understanding.

Yet the vows of this nation can only be fulfilled if we in this nation are first and, therefore, we intend to be first. In short, our leadership in science and in industry, our hopes for peace and security, our obligations to ourselves as well as others, all require us to make this effort, to solve these mysteries, to solve them for the good of all men, and to become the world's leading space-faring nation.

We set sail on this new sea because there is new knowledge to be gained, and new rights to be won, and they must be won and used for the progress of all people. For space science, like nuclear science and all technology, has no conscience of its own. Whether it will become a force for good or ill depends on man, and only if the United States occupies a position of pre-eminence can we help decide whether this new ocean will be a sea of peace or a new terrifying theater of war. I do not say that we should or will go unprotected against the hostile misuse of space any more than we go unprotected against the hostile use of land or sea, but I do say that space can be explored and mastered without feeding the fires of war, without repeating the mistakes that man has made in extending his writ around this globe of ours.

There is no strife, no prejudice, no national conflict in outer space, as yet. Its hazards are hostile to us all. Its conquest deserves the best of all mankind, and its opportunity for peaceful cooperation many never come again. But why, some say, the Moon? Why choose this as our goal? And they may well ask, Why climb the highest mountain? Why, 35 years ago, fly the Atlantic? Why does Rice play Texas?

We choose to go to the Moon. We choose to go to the Moon in this decade and do the other things, not because they are easy, but because they are hard, because that goal will serve to organize and measure the best of our energies and skills, because that challenge is one that we are willing to accept, one we are unwilling to postpone, and one which we intend to win, and the others, too.

It is for these reasons that I regard the decision last year to shift our efforts in space from low to high gear as among the most important decisions that will be made during my incumbency in the office of the presidency.

In the last 24 hours we have seen facilities now being created for the greatest and most complex exploration in man's history. We have felt the ground shake and the air shattered by the testing of a Saturn C-1 booster rocket, many times as powerful as the Atlas which launched John Glenn, generating power equivalent to 10,000 automobiles with their accelerators on the floor. We have seen the site where the F-1 rocket engines, each one as powerful as all eight engines of the Saturn combined, will be clustered together to make the advanced Saturn missile, assembled in a new building to be built at Cape Canaveral as tall as a 48-story structure, as wide as a city block, and as long as two lengths of this field.

Within these last 19 months, at least 45 satellites have circled the Earth. Some 40 of them were 'made in the United States of America' and they were far more sophisticated and supplied far more knowledge to the people of the world than those of the Soviet Union.

The Mariner spacecraft now on its way to Venus is the most intricate instrument in the history of space science. The accuracy of that shot is comparable to firing a missile from Cape Canaveral and dropping it in this stadium between the 40-yard lines.

Transit satellites are helping our ships at sea to steer a safer course. TIROS satellites have given us unprecedented warnings of hurricanes and storms, and will do the same for forest fires and icebergs. We have had our failures, but so have others, even if they do not admit them. And they may be less public. To be sure, we are behind, and will be behind for some time in manned flight. But we do not intend to stay behind, and in this decade, we shall make up and move ahead. The growth of our science and education will be enriched by new knowledge of our universe and environment, by new techniques of learning and mapping and observation, by new tools and computers for industry, medicine, the home, as well as the school. Technical institutions, such as Rice, will reap the harvest of these gains.

And finally, the space effort itself, while still in its infancy, has already created a great number of new companies, and tens of thousands of new jobs. Space and related industries are generating new demands in investment and skilled personnel, and this city and this state, and this region, will share greatly in this growth. What was once the furthest outpost on the old frontier of the West will be the furthest outpost on the new frontier of science and space. Houston, your city of Houston, with its Manned Spacecraft Center, will become the heart of a large scientific and engineering community. During the next five years the National Aeronautics and Space Administration expects to double the number of scientists and engineers in this area, to increase its outlays for salaries and expenses to $60 million a year; to invest some $200 million in plant and laboratory facilities; and to direct or contract for new space efforts over $1 billion from this center in this city.

To be sure, all this costs us all a good deal of money. This year's space budget is three times what it was in January 1961, and it is greater than the space budget of the previous eight years combined. That budget now stands at $5,400 million a year—a staggering sum, though somewhat less than we pay for cigarettes and cigars every year. Space expenditures will soon rise some more, from 40 cents per person per week to more than 50 cents a week for every man, woman, and child in the United States, for we have given this program a high national priority—even though I realize that this is in some measure an act of faith and vision, for we do not now know what benefits await us.

But if I were to say, my fellow citizens, that we shall send to the Moon, 240,000 miles away from the control station in Houston, a giant rocket more than 300 feet tall, the length of this football field, made of new metal alloys, some of which have not yet been invented, capable of standing heat and stresses several times more than have ever been experienced, fitted together with a precision better than the finest watch carrying all the equipment needed for propulsion, guidance, control, communications, food, and survival, on an untried mission, to an unknown celestial body, and then return it safely to earth, re-entering the atmosphere at speeds of over 25,000 miles per hour, causing heat about half that of the temperature of the Sun—almost as hot as it is here today—and do all this, and do it right, and do it first before this decade is out—then we must be bold.

I'm the one who is doing all the work, so we just want you to stay cool for a minute. [laughter]

However, I think we're going to do it, and I think that we must pay what needs to be paid. I don't think we ought to waste any money, but I think we ought to do the job. And this will be done in the decade of the sixties. It may be done while some of you are still here at school at this college and university. It will be done during the term of office of some of the people who sit here on this platform. But it will be done. And it will be done before the end of this decade.

I am delighted that this university is playing a part in putting a man on the Moon as part of a great national effort of the United States of America.

Many years ago, the great British explorer George Mallory, who was to die on Mount Everest, was asked why he wanted to climb it. He said, 'Because it is there.'

Well, space is there, and we're going to climb it, and the moon and the planets are there, and new hopes for knowledge and peace are there. And, therefore, as we set sail, we ask God's blessing on the most hazardous and dangerous and greatest adventure on which man has ever embarked.

The Mercury Missions

Only six months after its creation, NASA's first major task was the 1959 selection of the seven astronauts to prepare for Project Mercury. The program was designed to determine if humans could survive up in space and perform useful work there. Between 1961 and 1963, six astronauts ventured into space. The first two were sub-orbital, simply launching up above the atmosphere and landing in the Atlantic Ocean. During the other four missions, astronauts circled the Earth in orbit. They flew in a spacecraft so small that there was only room for one astronaut who would have to remain in a seated position during the entire flight.

Both of the rockets chosen for use in Project Mercury were originally designed for the United States military. The first two flights with an astronaut used the Redstone, while the four flights that orbited Earth launched atop a modified Atlas intercontinental ballistic missile.

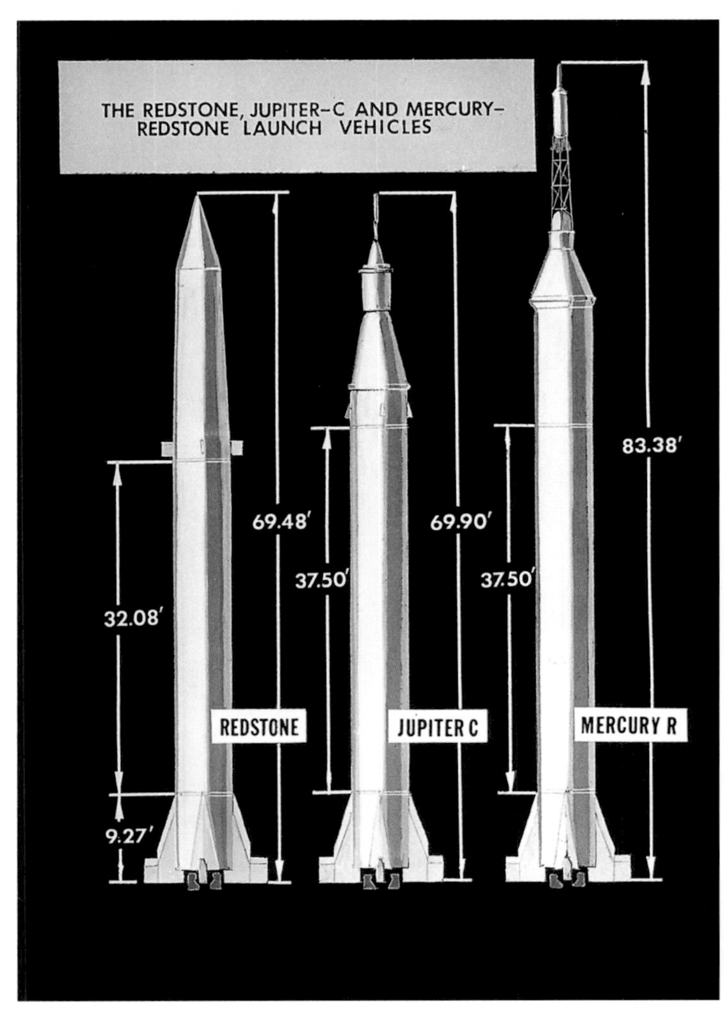

Above: This is a comparison illustration of the Redstone, Jupiter-C, and Mercury Redstone launch vehicles. The Redstone ballistic missile was a high-accuracy, liquid-propelled, surface-to-surface missile. Originally developed as a nose cone re-entry test vehicle for the Jupiter intermediate range ballistic missile, the Jupiter-C was a modification of the Redstone missile and successfully launched the first American Satellite, Explorer-1, in orbit on 31 January, 1958. The Mercury Redstone lifted off carrying the first American, astronaut Alan Shepard, in his Mercury spacecraft Freedom 7, on 5 May, 1961.

Top right: Launching of the Mercury-Redstone 3 (MR-3) spacecraft from Cape Canaveral on a suborbital mission on 5 May, 1961—the first US manned spaceflight.

[Photo and caption credit: NASA]

Bottom right: At the Cape Canaveral Air Force Base on 23 February, 1962 in Cocoa Beach, Florida, US President John F. Kennedy and astronaut John Glenn look into the Mercury Friendship 7 space capsule that carried Glenn into space. Glenn was awarded the Distinguished Service Medal by the president, following his historic flight. Those looking on include special assistant to the president Lawrence Larry O'Brien, Senator George Smathers of Florida, and director of the Manned Spacecraft Center Dr. Robert Gilruth.

[Photo and caption credit: NASA Image Collection / Alamy Stock Photo.]

The Mercury Astronauts

The project was named for the Roman winged god, Mercury. NASA chose seven astronauts for the program: Scott Carpenter, Gordon Cooper, John Glenn, Gus Grissom, Wally Schirra, Alan Shepard, and Deke Slayton. All would fly Mercury missions with the exception of Slayton, who was grounded from flying due to an irregular heart rhythm. He finally flew into space during the Apollo-Soyuz Test Project mission in 1975.

Above: Primate chimpanzee "Ham" is fitted into the couch of the Mercury-Redstone 2 (MR-2) capsule #5 prior to its test flight, which was conducted on 31 January 1961.

[Photo and caption credit: NASA]

Opposite: Project Mercury astronauts, whose selection was announced on 9 April 1959, only six months after the National Aeronautics and Space Administration was formally established on 1 October 1958. Front row, left to right: Walter M. Schirra, Jr., Donald K. Slayton, John H. Glenn, Jr., and M. Scott Carpenter; back row, Alan B. Shepard, Jr., Virgil I. "Gus" Grissom and L. Gordon Cooper.

[Photo and caption credit: NASA Image Collection / Alamy Stock Photo]

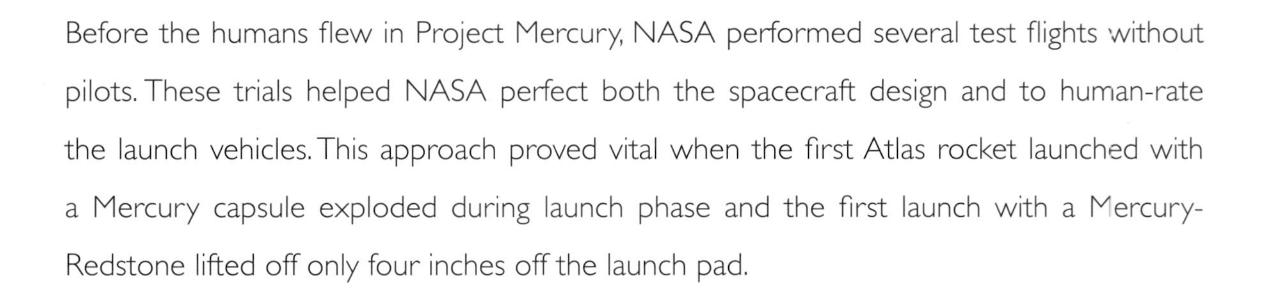

Before the humans flew in Project Mercury, NASA performed several test flights without pilots. These trials helped NASA perfect both the spacecraft design and to human-rate the launch vehicles. This approach proved vital when the first Atlas rocket launched with a Mercury capsule exploded during launch phase and the first launch with a Mercury-Redstone lifted off only four inches off the launch pad.

Additionally, prior to sending humans into space, NASA chose four other "astronauts" to fly as part of the Mercury program. Two rhesus monkeys named Sam and Miss Sam, along with two chimpanzees, Ham and Enos, did their part to ensure safe spaceflight. Sam and Miss Sam made short flights launched by Little Joe rockets to test the Mercury spacecraft launch abort system. Ham flew on a Redstone rocket for a sub-orbital mission into space. Enos launched in a Mercury spacecraft lifting off atop an Atlas rocket making two orbits. With all four primates making it home safely, NASA was ready to send humans to space, with the Mercury astronauts going on to successfully complete their missions.

Alan Shepard and Freedom 7

Alan Shepard made history when he became the first American to fly in space on 5 May 1961. The mission only lasted 15 minutes, going up into space before coming back down and landing in the Atlantic Ocean. During Project Mercury, each astronaut named his spacecraft; Shepard was the first, dubbing his space capsule Freedom 7, with a view to the competition between the USA and the Soviets. Ten years later, Shepard walked on the Moon as commander of the Apollo 14 mission making him the only Mercury astronaut to do so.

Right: The recovery of the Freedom 7 (MR-3) capsule by a US Marine helicopter.
Opposite: Shepard walks away from his Freedom 7 capsule following a post-flight inspection aboard the carrier USS Champlain.
[Photo and caption credit: NASA Image Collection / Alamy Stock Photo]

LAKE CHAMPLAIN

"There are people who make things happen, there are people who watch things happen, and there are people who wonder what happened. To be successful, you need to be a person who makes things happen."

Commander James Lovell, Apollo 13

Gus Grissom and Liberty Bell 7

The second astronaut to fly was Gus Grissom with a sub-orbital mission on 21 July 1961. Keeping with the concept of "freedom," he named his capsule Liberty Bell 7. He made a successful 16-minute sub-orbital flight. However, after splashing down in the Atlantic Ocean, the spacecraft's hatch separated prematurely causing the capsule to fill with water. Grissom was able to jump out and was rescued by a Marine helicopter crew. Despite a valiant attempt, another helicopter crew failed to lift the now heavy Liberty Bell 7 from the water, and it quickly sank to a depth of three-miles to the Atlantic Ocean floor.

Three years later, Grissom was command pilot for the first Gemini mission. Sadly, he died in 1967 along with two crewmates in a flash fire that broke out in an Apollo spacecraft during a countdown simulation on a launch pad. After 38 years, an expedition sponsored by the Discovery Channel began an underwater salvage activity to recover Liberty Bell 7. Curt Newport, an expert in such operations, led the expedition. The team found and successfully raised the capsule and brought it to Port Canaveral. Liberty Bell 7 was later moved to the Kansas Cosmosphere and Space Center in Hutchinson, Kansas, where it was restored and is now on public display.

Below: Donning a spacesuit for the Mercury-Redstone 4 (MR-4) mission, astronaut Virgil I. (Gus) Grissom chats with spaceflight equipment specialist Joe W. Schmidt in the personal equipment room of Hanger S at Cape Canaveral, Florida. Shortly after this photograph was taken, the launch was postponed two days due to unfavorable weather conditions in the area.

[Photo and caption credit: NASA]

John Glenn and Friendship 7

John Glenn was next in line, but this time he would be going all the way to Earth orbit. On 20 February, 1962, he became the first American to do so aboard his capsule Friendship 7. Next up and the second to orbit Earth, was Scott Carpenter whose capsule was Aurora 7. This would be his only trip into space. Both Glenn and Carpenter completed three orbits.

Right: American astronaut John Herschel Glenn Jr. looking into a space globe known as a "Celestial Training Device" at the Aeromedical Laboratory at Cape Canaveral, Florida. Glenn is shown here in February 1962, the same month he became the first American to orbit Earth on NASA's Mercury-Atlas 6 Mission.

[Photo and caption credit: NASA Image Collection / Alamy Stock Photo]

"The most important thing we can do is inspire young minds and to advance the kind of science, math, and technology education that will help youngsters take us to the next phase of space travel."

Astronaut John Glenn

Scott Carpenter and Aurora 7

Right: President John F. Kennedy stands with astronaut Lieutenant Commander M. Scott Carpenter (center right), during Lt. Cdr. Carpenter's visit to the White House following his Mercury-Atlas 7 orbital flight (also known as Aurora 7) on 24 May 1962.

Opposite: Astronaut M. Scott Carpenter, prime pilot for the Mercury-Atlas 7 (MA-7) flight, is seen in Hanger S crew quarters during a suiting exercise.

[Photos and captions credit: NASA Image Collection / Alamy Stock Photo]

Below: Astronaut Walter M. Schirra (pronounced Shuh-RAH) climbs into his Sigma 7 spacecraft, becoming the fifth American to go up in space as part of NASA's Mercury program. He orbited the Earth six times in the Sigma 7 spacecraft in a nine-hour flight focused mainly on technical evaluation rather than on scientific experimentation. This was the longest US-manned orbital flight yet achieved in the Space Race, though well behind the several-day record set by the Soviet Vostok 3 earlier in the year. It confirmed the Mercury spacecraft's durability ahead of the one-day Mercury-Atlas 9 mission that followed in 1963.
[Photo and caption credit: NASA Image Collection / Alamy Stock Photo]

Opposite: Launch of the Mercury-Atlas 8 "Sigma 7" mission on 3 October 1962.
[Photos and captions credit: NASA]

Wally Schirra and Sigma 7

Astronaut Wally Schirra made the fifth Mercury flight aboard the spacecraft he named Sigma 7. He doubled the two previous Mercury missions by orbiting the Earth six times. Gordon Cooper flew on the final Mercury mission on 15–16 May 1963, spending 34 hours circling Earth 22 times aboard the capsule he named Faith 7.

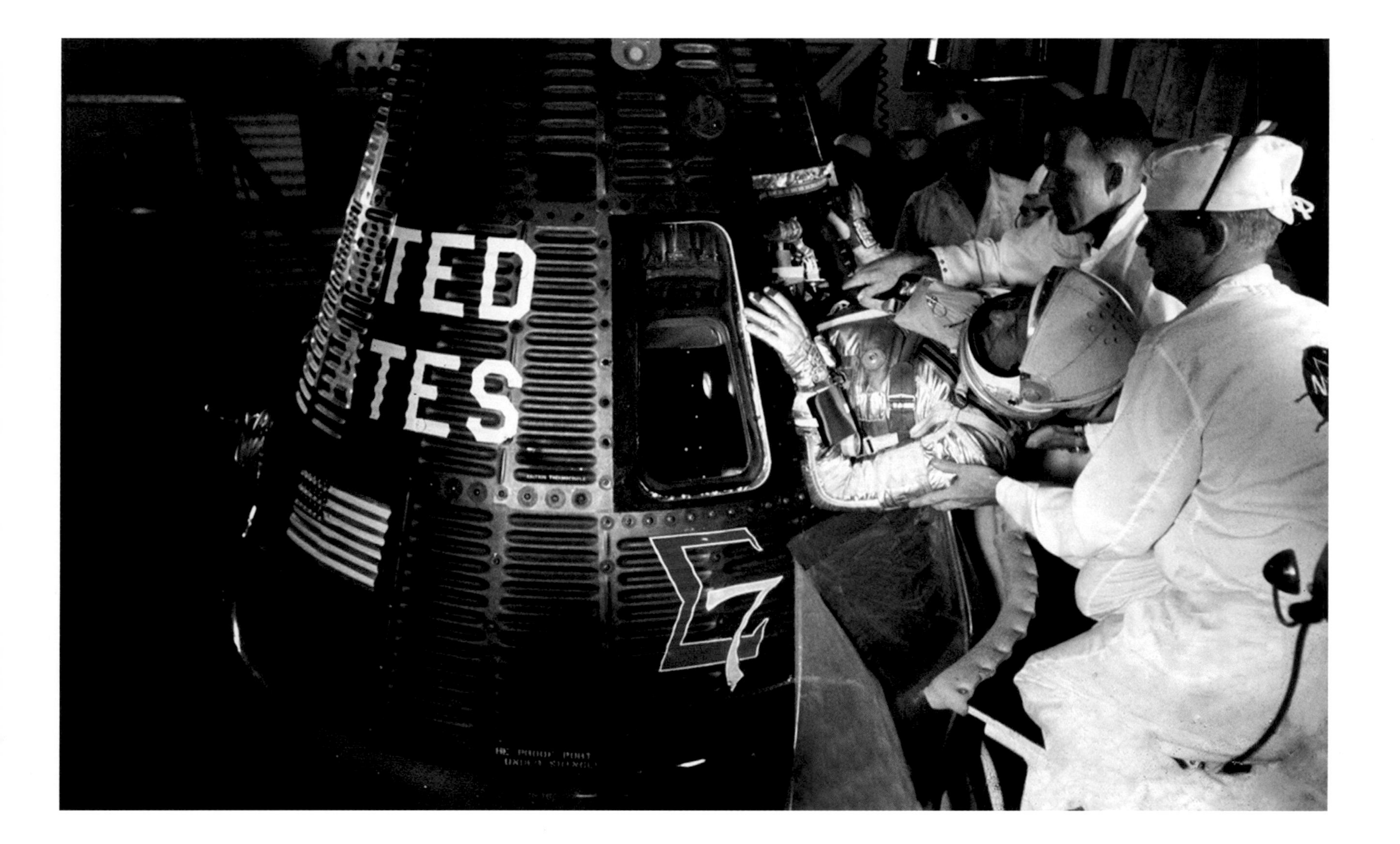

Gordon Cooper and Faith 7

Right: Official portrait of astronaut L. Gordon Cooper for the NASA Mercury-Atlas 9 "Faith 7" spacecraft mission. Cooper was the sixth and final man, flying the longest Mercury mission on 15 May 1963.

[Photo and caption credit: NASA Image Collection / Alamy Stock Photo]

Opposite: Preparation for launch of Mercury-Atlas 9 (MA-9).

[Photo and caption credit: NASA]

"Father, we thank you, especially for letting me fly this flight, for the privilege of being able to be in this position, to be in this wondrous place, seeing all these many startling, wonderful things that you have created."

Astronaut Leroy Gordon "Gordo" Cooper, Jr.

Left: Astronaut L. Gordon Cooper Jr., prime pilot for the Mercury-Atlas 9 (MA-9) mission, is pictured just after his helmet had been removed. He had just spent approximately five hours in the spacecraft during altitude chamber tests.

[Photo and caption credit: NASA Image Collection / Alamy Stock Photo]

Human Computers

Mary Jackson, Katherine Johnson, and Dorothy Vaughan were highly skilled mathematicians who worked for NASA performing calculations to support advances in aviation and spaceflight. Johnson calculated the trajectory for America's first trip into space with Shepard. After computers were used to complete similar calculations for John Glenn's orbital flight, he insisted Johnson verify the accuracy of the machine's data for his mission.

Above: Mary Winston Jackson (1921–2005) was an African-American mathematician and aerospace engineer at the National Advisory Committee for Aeronautics (NACA), which in 1958 became the National Aeronautics and Space Administration (NASA). Mary worked at Langley Research Center in Hampton, Virginia, for most of her career, starting as a "computer" at the segregated West Area Computing division. She took advanced engineering classes and in 1958 became NASA's first black female engineer. Jackson was featured in the movie *Hidden Figures* (2016), as well as in the book upon which the film was based.

Above: African-American mathematician, Dorothy Vaughan (1910–2008), was one of the super "human computers" working at NACA, which later became NASA.
Opposite: Katherine Coleman Goble Johnson (1918–2020), an African-American mathematician who made contributions to the United States' aeronautics and space programs with the early application of digital electronic computers at NASA.

[Phcto and caption credits: Alamy Stock Photo]

The Importance of Project Mercury

NASA's Mercury program was a crucial first step for the fledgling space agency. The six missions confirmed that humans could survive flights into space and perform useful work while there. This set the stage of the more complex missions of Project Gemini and meeting President Kennedy's goal of reaching the Moon.

Above: President John F. Kennedy receives a gift of an American flag from astronaut Lieutenant Colonel John H. Glenn, Jr. (right); Lt. Col. Glenn carried the flag in his space suit during his orbital flight aboard Mercury-Atlas 6, also known as Friendship 7. Special assistant to the president, Kenneth P. O'Donnell, stands in the background. Oval Office, White House, Washington, D.C.

[Photo and caption credit: American Photo Archive / Alamy Stock Photo]

Above: President Kennedy and Vice President Johnson watching the lift-off of the first American in space on 5 May 1961. Left to right: Vice President Johnson, Arthur Schlesinger, Adm. Arleigh Burke, President Kennedy, Mrs. Kennedy. White House, office of the president's secretary.

[Photo and caption credit: Cecil Stoughton/The White House / Alamy Stock Photo]

The Gemini Program

Following the successful Mercury missions, NASA launched Project Gemini, a crucial bridge between Mercury and the Apollo Program, preparing to meet President Kennedy's goal of landing the first humans on the Moon. Named after the constellation Gemini, Latin for twins, the Gemini spacecraft was larger, with the capsule designed for two people instead of one. This enabled the agency to learn more lessons about space travel, such as changing orbits, rendezvous, docking, and working outside a spacecraft. The Gemini had many advances in technology, for example, where the Mercury spacecraft could only redirect where it was pointed in space, Gemini could change its orbit, maneuver, and rendezvous with another spacecraft.

The Gemini capsule launched on the two-stage Titan II rocket originally designed as an intercontinental ballistic missile. The rocket was modified and, like the Redstone and Atlas, man-rated to make it safe for astronauts. The Gemini I and II missions were launched on the rocket without a crew, to test its safety and confirm reliability. Ten crews flew missions between 1965 and 1966.

The Gemini Missions

The Gemini missions proved NASA had mastered the technology required for the Apollo lunar missions. The first flight with a crew was Gemini III. That flight confirmed the new vehicle could alter its orbit as planned.

The Gemini IV mission lasted four days and included the first spacewalk by an American. Gemini V set a new space endurance record staying in orbit for eight days. Gemini VI and VII missions orbited at the same time and met each other in orbit, completing the first rendezvous in space. Additionally, the Gemini VII astronauts remained in space for 14 days, the length of the longest lunar missions planned for Apollo.

Gemini VIII was the first to dock with another spacecraft in orbit. However, NASA and the Gemini VIII crew demonstrated how quick thinking was crucial in human spaceflight. When a thruster on the capsule misfired, the spacecraft briefly spun wildly. Neil Armstrong and David Scott were able to regain control and made an emergency landing in the western Pacific Ocean.

The Gemini IX mission performed different methods for completing a rendezvous with another orbiting spacecraft. It also included a spacewalk. Gemini X docked with another spacecraft, an Agena, and used its engine to propel both vehicles to a higher orbit. The Gemini XI crew used an Agena to fly to an altitude of 850 miles, higher than any prior space mission. Gemini XII concluded the program solving problems encountered during earlier spacewalks.

While the Mercury missions proved astronauts could fly in space, the Gemini program gave NASA much of the experience needed to land a human on the Moon. NASA learned what happens when astronauts spend up to two weeks in space and how best to work outside a spacecraft. The agency also determined how to dock two spacecraft together in space. Going to the Moon would require all these skills and more. By the end of the program's final mission, Gemini proved NASA was ready for Apollo.

Gemini 1

Top Left: Gemini 1 was the first mission in NASA's Gemini program and took place on 8 April 1964. An uncrewed test flight of the Gemini spacecraft, its main objectives were to test the structural integrity of the new spacecraft and the modified Titan II launch vehicle. The spacecraft stayed attached to the second stage of the rocket.

Gemini 2

Bottom Left: On 19 January 1965, the unmanned Gemini 2 flight launched. The second Titan II Gemini Launch Vehicle (GLV-2) carried the unmanned, instrumented Gemini spacecraft (GT-2) for a suborbital shot preliminary to the first US two-man Gemini mission.

[Photo and caption credit: NASA]

Gemini 3

Above: Gemini 3 was the first spacecraft to maneuver in orbit. On 23 March 1965, Virgil Grissom and John Young carried out the mission.

[Photo and caption credit: NASA Image Collection / Alamy Stock Photo]

Gemini 4

Above: On 3 June 1965, NASA astronaut Ed White floated in the microgravity of space outside the Gemini 4 spacecraft with the blue of the Earth below. This was the first spacewalk by an American in Earth's orbit.

[Photo and caption credit: NASA Image Collection / Alamy Stock Photo]

"Do not go where the
path may lead, go instead
where there is no path
and leave a trail."

Ralph Waldo Emerson

Gemini 5

Left: NASA launched the Gemini-5 spacecraft from Pad 19 on 21 August 1965 with astronaut and command pilot Gordon Cooper Jr. and astronaut and pilot Charles Conrad Jr. The planned eight-day orbital mission was the longest manned spaceflight at that time.

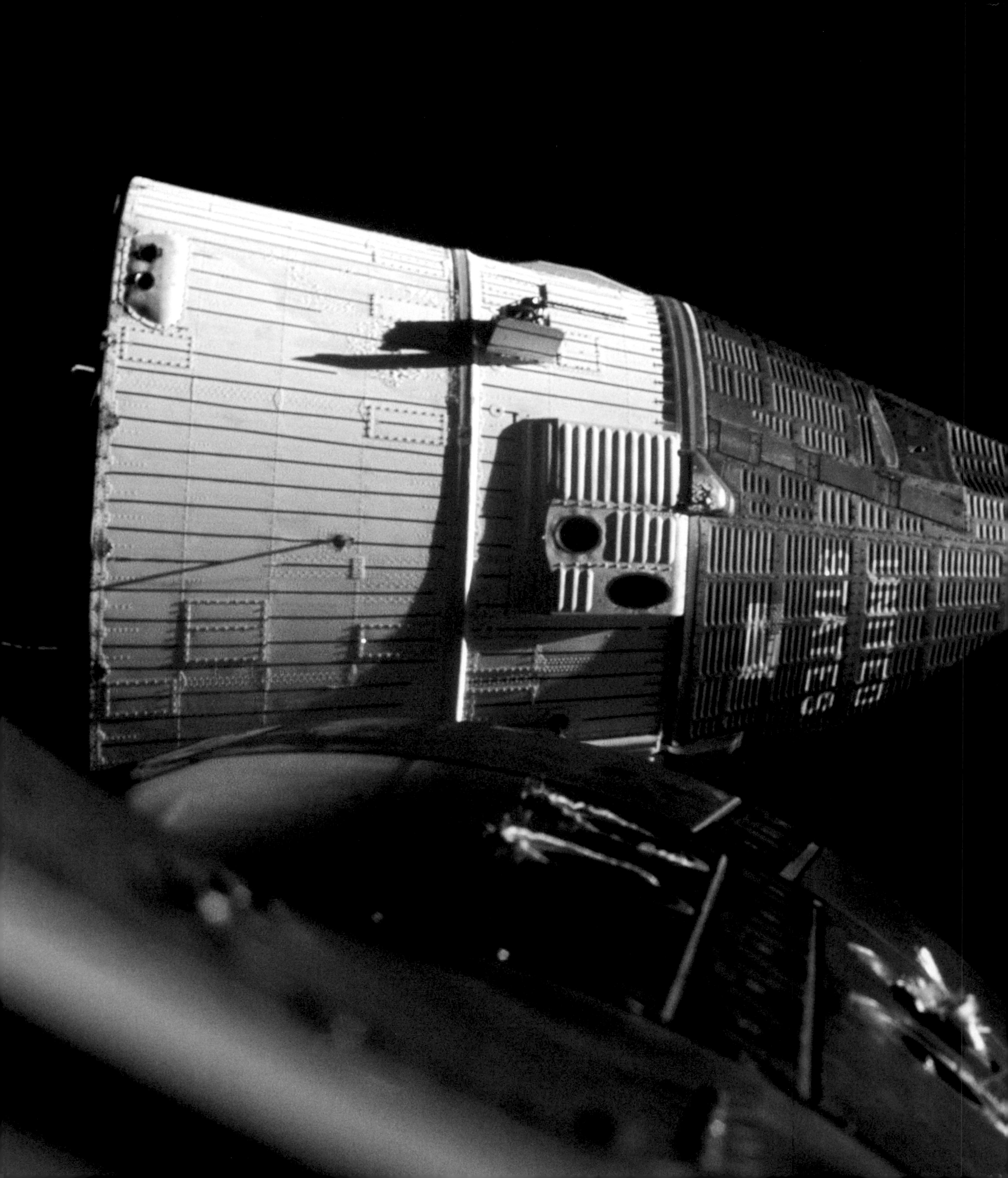

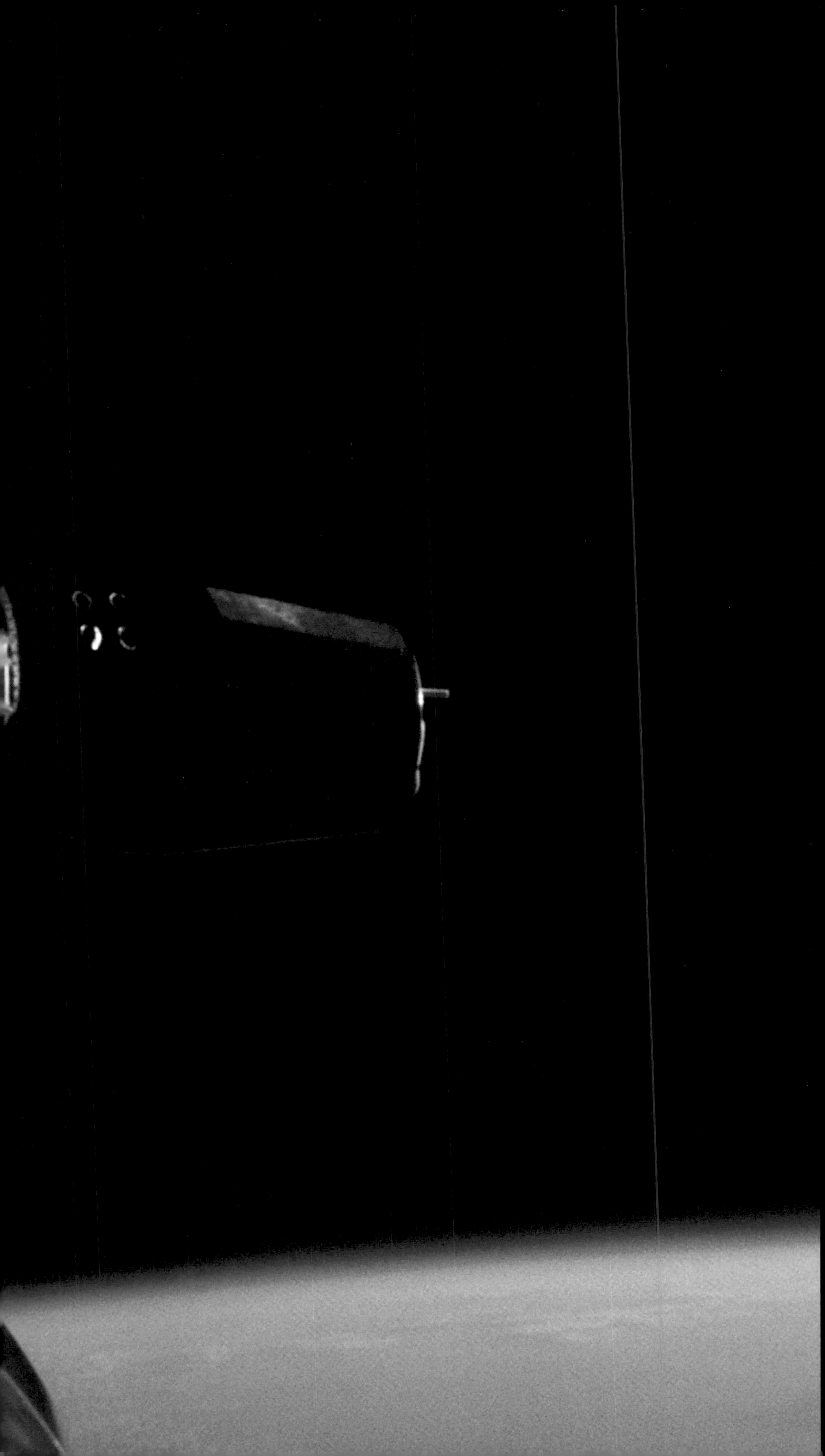

Gemini 6A

Left: Gemini 6A was scheduled to launch on 12 December 1965, but the launch was aborted one second after engine ignition because an electrical umbilical separated prematurely. This was the first time an astronaut mission was aborted after ignition start. The mission launched successfully from Complex 19 on 15 December.

This photograph of the NASA's Gemini-7 spacecraft was taken on 15 December 1965 through the hatch window of the Gemini-6 spacecraft during rendezvous and station-keeping maneuvers, at an altitude of approximately 160 miles. The photograph was taken with a Hasselblad camera using Kodak SO 217 film with an ASA of 1964.

Gemini 7

Opposite: This photograph of the Gemini-Titan 7 (GT-7) was taken from the Gemini-Titan 6 (GT-6) during the historic rendezvous of the two spacecraft on 15 December 1965. The spacecrafts are some 37 feet apart here and Earth can be seen below. Command pilot Walter M. Schirra Jr. and pilot Thomas P. Stafford were inside the GT-6 spacecraft, while crewmen for the GT-7 mission were command pilot Frank Borman and pilot James A. Lovell Jr. Borman and Lovell set a new space endurance record of 13 days and 19 hours.

[Photo and caption credit: NASA]

Gemini 8

Above: Gemini 8 was the first docking of two spacecraft which took place on 16–17 March 1966. These two astronauts are command pilot Neil A. Armstrong (left) and pilot David R. Scott, the Gemini-8 prime crew.

[Photo credit: NASA Image Collection / Alamy Stock Photo]

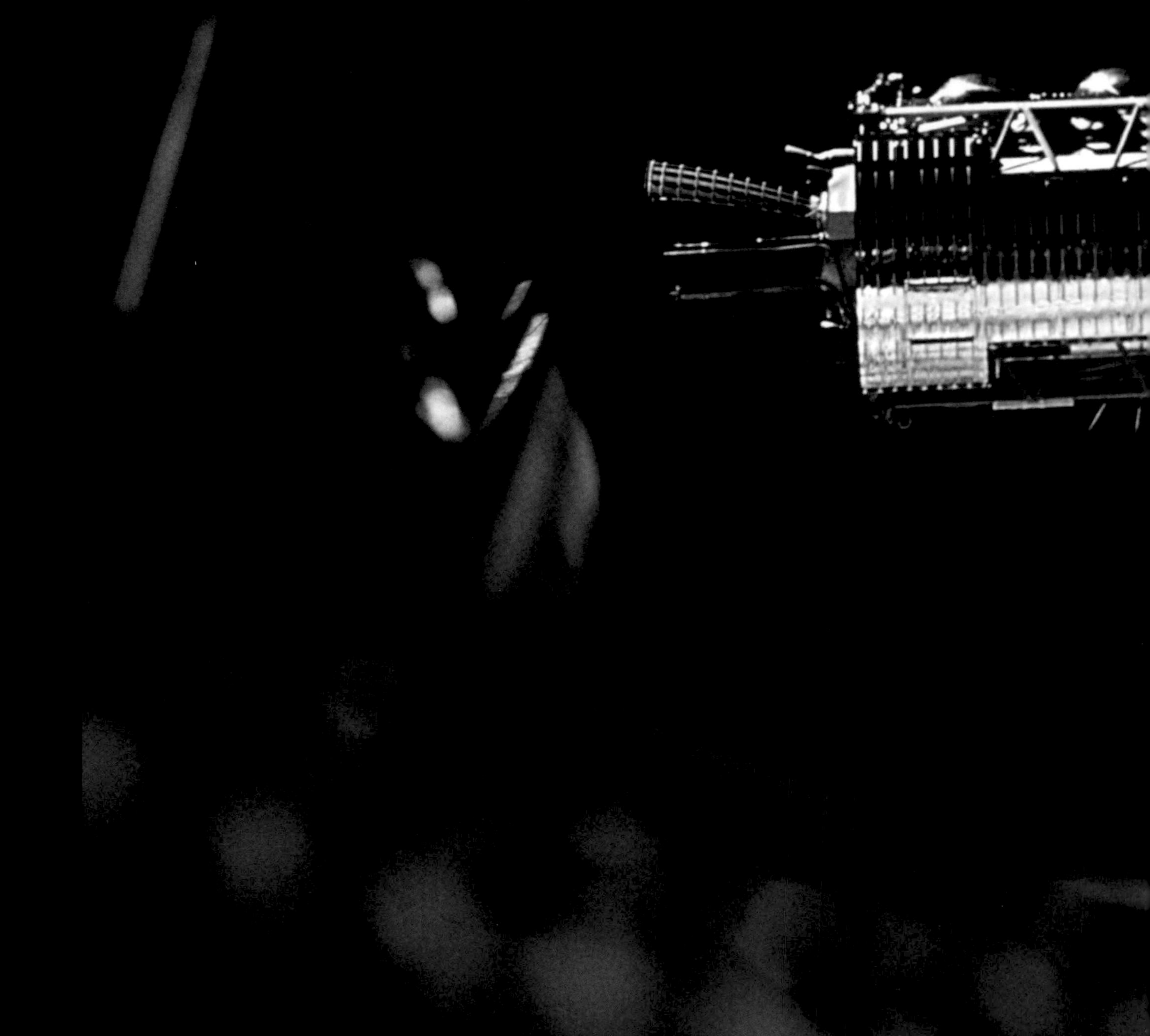

Right: Closer view of the Agena Target Docking vehicle seen from the Gemini-8 spacecraft during rendezvous in space on 16 March 1966.

[Photo and caption credit: NASA]

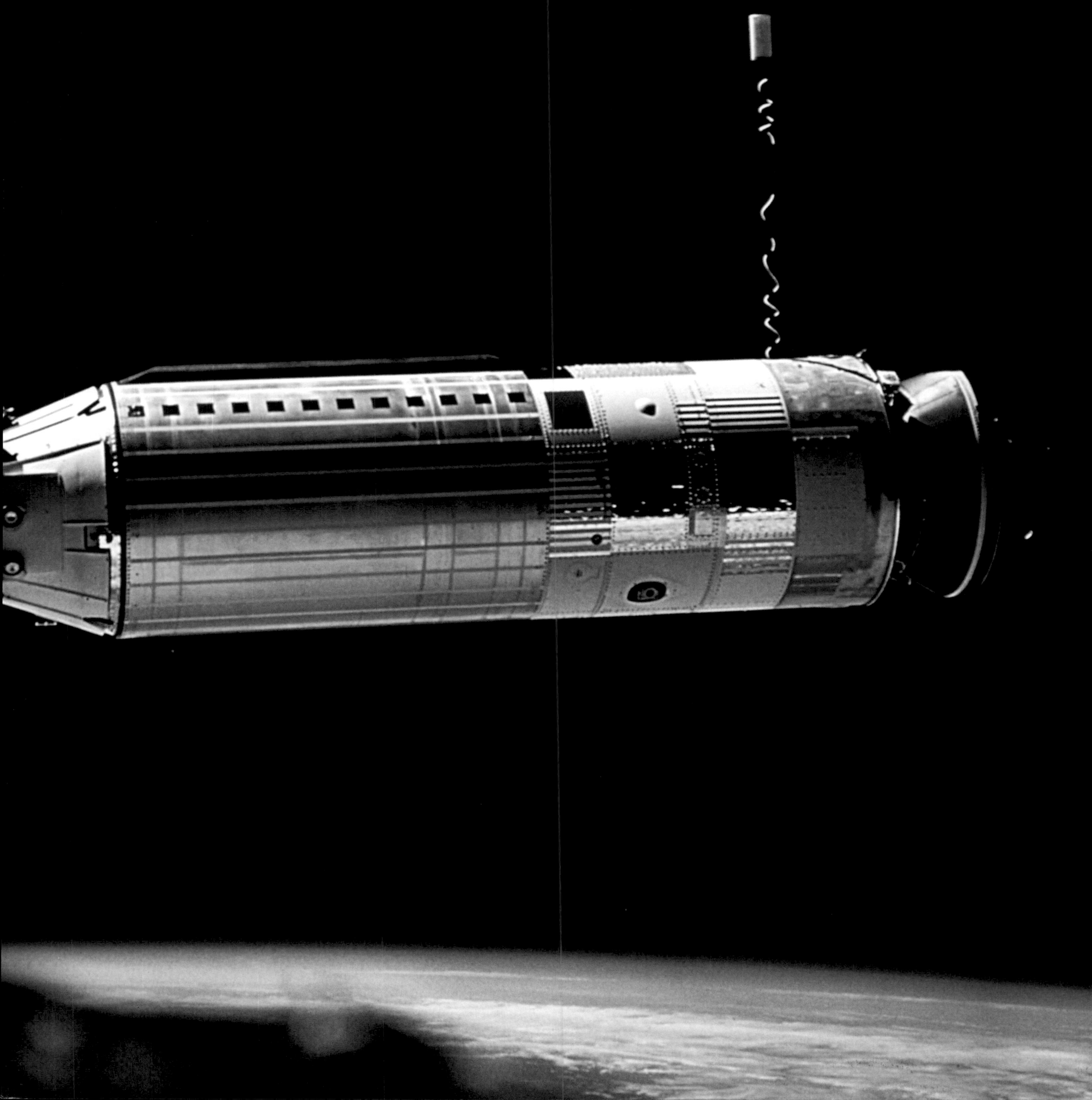

LEFT

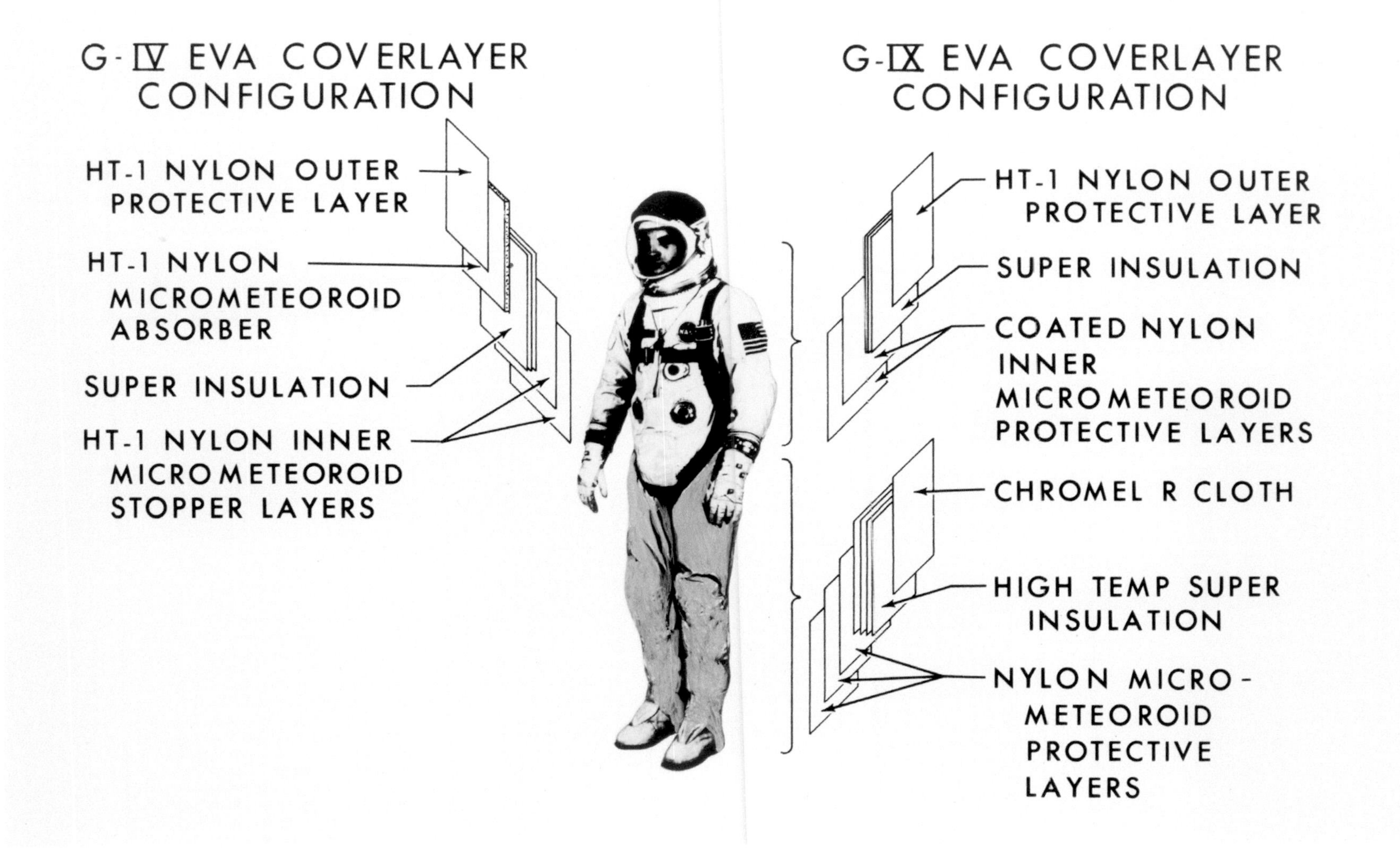

Gemini 9

Above: Drawing of the Gemini-9 extravehicular spacesuit, with a comparison of the breakdown between the Gemini-4 suit layers and the Gemini-9 suit layers.

[Photo and caption credit: NASA]

Opposite: Astronaut Thomas P. Stafford, command pilot of the Gemini-9A spaceflight, is photographed during the Gemini-9A mission inside the spacecraft.

[Photo and caption credit: The NASA Library / Alamy Stock Photo]

Gemini 10

Above: Close-up of astronaut Michael Collins, Gemini-10 pilot, making final adjustments and checks in the Gemini spacecraft during prelaunch countdown. In the background, to the right, is command pilot John W. Young.

[Photo and caption credit: NASA]

Opposite: The Gemini-10 spacecraft is launched from Complex 19 at 5.20 p.m. on 18 July 1966. A time exposure creates the illusion of multiple rocker arms. Onboard are command pilot John W. Young and pilot Michael Collins, The Gemini 10 mission was the first space walk from one spacecraft to another.

[Photo and caption credit: NASA]

Gemini 11

Above: The Agena Target Docking Vehicle is tethered to the Gemini-11 spacecraft during its 31st revolution of Earth. Area below is the Gulf of California and Baja California at La Paz. Taken on 14 September 1966 with a J.A. Maurer 70mm camera, using Eastman Kodak, Ektachrome, MS (S.O. 368) color film.

[Photo and caption credit: NASA]

Gemini 12

Above: Excellent stereo and side view of the Agena Target Docking Vehicle as seen from the Gemini-12 spacecraft during rendezvous and docking mission in space. The two spacecraft are 50 feet apart.

Opposite: The Gemini 12 astronauts, James Lovell and Edwin Aldrin, lifted off aboard a Titan launch vehicle from the Kennedy Space Center on 11 November 1966, an hour and a half after their Agena target vehicle was orbited by an Atlas rocket. Launched atop an Atlas booster, the Agena target vehicle was a spacecraft used by NASA to develop and practice orbital space rendezvous and docking techniques in preparation for the Apollo program lunar missions. The objective was for Agena and Gemini to rendezvous in space and practice docking procedures. An intermediate step between Project Mercury and the Apollo Program, the Gemin Program's major objectives were to subject two men and supporting equipment to long duration flights; and to perfect rendezvous and docking with other orbiting vehicles, methods of re-entry, and landing of the spacecraft. Three space walks by Aldrin during this mission solved problems of exhaustion and suit overheating that had occurred during previous flights.

UNITED STATES

When, during the late 1960s and early '70s, technology consisted of the first inexpensive transistor hand-held radios and over-priced digital wristwatches, seeing man land on the Moon was really out-of-this-world! I fondly recall staring up at the Moon and pondering that an astronaut was actually driving a lunar Moon rover, doing donuts and popping wheelies on the surface of the Moon. The grainy television coverage in black and white also added a bit of a 'horror film' spooky vibe, too. However, it was viewing the return from the Moon that was thrilling; from the communication black out when the orbiter circled the back of the Moon; to the earthbound cameras searching the skies to find the plummeting capsule nearly burning up; to when the first parachute slowed the capsule's descent; to the splashdown and astronaut recovery. There was nothing else on television that measured up to the Apollo missions. Without a doubt, space exploration during this period was a unifier since the vast majority of the population joined together in their excitment and national pride.

NASA

"That's one small step for man, one giant leap for mankind."

Neil Armstrong

Landing Man on the Moon

On 20 July 1969, President John F. Kennedy's goal was achieved when Apollo 11 commander Neil Armstrong stepped off the lunar module's ladder and onto the Moon's surface. "That's one small step for man, one giant leap for mankind," Armstrong said. And the world, indeed, watched in awe.

The American effort to place Armstrong on the Moon, to speak those words and fulfil a long-standing dream, went beyond landing on the lunar surface and returning safely to Earth. NASA had developed technologies that achieved goals, such as carrying out a program of scientific exploration of the Moon and developing what it took to work there. There was also the desire for the United States to achieve pre-eminence in the "Space Race" over their Cold War adversary, the Soviet Union.

Among the flight modes considered for landing humans on the Moon, was one which involved sending an enormous rocket, called Nova, directly to the Moon. Another required assembly of a similar spacecraft in Earth orbit with multiple launches of smaller Saturn 1B rockets. The approach selected in 1962 was called, "lunar orbit rendezvous." This plan involved missions that launched a single Saturn V rocket with two spacecraft. The command module was made up of the crew quarters and flight-control section, along with a service module containing the propulsion and spacecraft support systems. The separate lunar module was designed to take two of the three crewmembers to the lunar surface, support them on the Moon and return them to rendezvous with the command/service module in lunar orbit. Ultimately, NASA used the Saturn 1B rocket for Earth orbital flights and the Saturn V for lunar missions.

Opposite: Apollo 1 mission badge, designed by Allen Stevens of North American Rockwell in 1966 for what was to be the first manned Apollo mission.

[Photo and caption credit: NASA]

Apollo 1 Tragedy

The first Apollo mission, a flight with Gus Grissom, Ed White, and Roger Chaffee, was scheduled to launch in February 1967. But as they were conducting a countdown simulation on a launch pad at Cape Kennedy on 27 January, a flash fire broke out in their spacecraft. In the 100 percent oxygen atmosphere used by NASA at the time, the fire spread quickly and killed all three men.

There was an exhaustive investigation into the causes of the fire and an extensive redesign of the Apollo spacecraft. While NASA officials made required modifications, it would be more than a year before astronauts would fly. The Saturn 1B launch vehicle originally assigned to launch the first crewed Apollo would later be used to launch a lunar module into orbit on an uncrewed test flight.

Later, NASA's associate administrator for manned space flight, Dr George E. Mueller, announced some of the plans going forward. The mission originally scheduled with Grissom, White, and Chaffee would be known as Apollo 1. He added that the first Saturn V launch, scheduled for November 1967, would be Apollo 4. The test flight of the lunar module was designated as Apollo 5.

Above: Closeup view of the interior of Apollo launch pad, showing the effects of the intense heat of the flash fire which killed the prime crew. Cape Kennedy, Florida.
[Photo and caption credit: NASA]

Above: Apollo 1 astronauts (left to right) Gus Grissom, Ed White, and Roger Chaffee, pose in front of Launch Complex 34 which is housing their Saturn 1 launch vehicle. The astronauts later died in a fire on the pad.

Apollo 4 and the First Saturn V

On 9 November 1967, Apollo 4 successfully launched on the first uncrewed flight to test the Saturn V launch vehicle that would eventually send astronauts to the Moon. The second took place early on 4 April 1968. Apollo 6 was a qualified success, even though two first-stage engines shut down prematurely and the third-stage engine failed to reignite after reaching orbit.

Apollo 7

On 11 October 1968, Apollo 7 launched the first crewed mission of the Apollo Program to Earth orbit. During the 11-day flight, the astronauts tested the spacecraft systems and conducted the first live television broadcasts from an American spacecraft. All three crewmembers, Wally Schirra, Walt Cunningham, and Donn Eisele, developed head colds during their time in space but they completed their mission objectives, exhibiting both resilience and adaptability.

Opposite: Early morning view of Pad A, Launch Complex 39, at the Kennedy Space Center, showing the 363-foot tall Apollo 4/Saturn V space vehicle ready for launch.

[Photo and caption credit: NASA]

Above: The morning sun reflects on the Gulf of Mexico and the Atlantic Ocean as seen from the Apollo 7 spacecraft at an altitude of 120 nautical miles above Earth. Most of Florida peninsula appears as a dark silhouette. This photograph was made during the spacecraft's 134th revolution of Earth, some 213 hours and 19 minutes after lift-off.

[Photo and caption credit: NASA]

Opposite: North American Rockwell Corporation artist's concept depicting the Apollo command module, oriented in a blunt and forward attitude, re-entering Earth's atmosphere after returning from a trip to the Moon. Cutaway view illustrates position of the three astronauts in the command module.

[Picture and caption credit: NASA]

Apollo 8

The success of Apollo 7, delays in the development of the lunar module, and concerns that the Soviet Union was about to launch a cosmonaut to fly a mission to circle around the Moon, led NASA to change the flight plan for the next mission. Apollo 8 was switched from a mission to circle the Earth at a high altitude, to a historic mission to orbit the Moon. Frank Borman, Jim Lovell, and Bill Anders were the first crew to launch atop the powerful Saturn V rocket. They spent 20 hours circling the Moon ten times. During a memorable television broadcast on Christmas Eve 1968, the crew read from the Bible's book of Genesis with the largest TV audience ever at the time. Their images and words were seen and heard by an estimated one billion people in 64 countries.

Apollo 8 also provided NASA with an opportunity to test the Apollo command module's systems, including communications, tracking, and life-support during the trip to the Moon and evaluate crew performance in lunar orbit. They photographed much of the lunar surface, including both the far and near sides. The imagery provided information on landmarks as well as other scientific information necessary for future Apollo landings. Anders also took what has come to be known as the iconic "Earthrise" photograph.

Right: North American Rockwell artist's concept illustrating a phase of the scheduled Apollo 8 lunar orbit mission. Here, the Apollo 8 spacecraft lunar module adapter (SLA) panels, which have supported the command and service modules, are jettisoned, separating the spacecraft from the SLA and deploying the high gain (deep space) antenna. This is done by astronauts firing the service module reaction control engines.

[Photo and caption credit: NASA]

Above: The rising Earth is about five degrees above the lunar horizon in this telephoto view taken from the Apollo 8 spacecraft near 110 degrees east longitude. The horizon, about 570 kilometers (350 statute miles) from the spacecraft, is near the eastern limb of the Moon as viewed from Earth. Width of the view at the horizon is about 150 kilometers (95 statute miles). On Earth 240,000 statute miles away the sunset terminator crosses Africa. The crew took the photo on the morning of 24 December 1968. The South Pole is in the white area near the left end of the terminator. North and South America are under the clouds.

[Photo and caption credit: NASA]

Apollo 9

With the command/service module proven in both Earth orbit and around the Moon, the lunar module was ready for its first crewed test. During the Apollo 9 mission in March 1969, commander Jim McDivitt and lunar module pilot Rusty Schweickart flew the lander for six hours while separated from command module pilot David Scott. The lunar module successfully demonstrated it could fly independently and rendezvous and dock with the command module. Additionally, Schweickart climbed out of the lunar module to perform a spacewalk standing on the lunar module's "front porch." This verified that the spacesuit astronauts would wear on the Moon worked as planned.

Opposite: North American Rockwell artist's concept illustrating a part of the planned Apollo 9 extravehicular activity on the fourth day of the mission as the command and service modules are docked to the lunar module.

Above: A view of the Apollo 9 lunar module "Spider" in a lunar landing configuration, as photographed from the command and service modules on 7 March 1969, the fifth day of the Apollo 9 Earth-orbital mission.

[Photo and caption credits: NASA]

Apollo 10

After Apollo 9 tested the lunar module in Earth orbit, the next flight was conducted around the Moon. Apollo 10 served a "dress-rehearsal" for the first lunar landing and was highlighted by the first color television images broadcast from space. Except for the actual touch down on the surface, the crew tested most aspects of the mission to land. Mission commander Tom Stafford and lunar module pilot Gene Cernan flew the lunar module for eight hours, coming within 10 miles of the lunar surface. In doing so, they passed over the Sea of Tranquillity, the site targeted for Apollo 11.

Opposite: A view of Earth from 36,000 nautical miles away as photographed from the Apollo 10 spacecraft during its trans-lunar journey toward the Moon. While the Yucatan Peninsula is obscured by clouds, nearly all of Mexico north of the Isthmus of Tehuantepec can be clearly delineated. The Gulf of California and Baja California and the San Joaquin Valley can be easily identified. Also, the delta of the Rio Grande River and the Texas coast are visible. Note the color differences (greens – east, browns – west) along the 100 degrees meridian. The crew members on Apollo 10 are commander Thomas P. Stafford, command module pilot John W. Young, and lunar module pilot Eugene E. Cernan. Astronaut Young remained in lunar orbit, in the command and service module's "Charlie Brown", while astronauts Stafford and Cernan descended to within nine miles of the lunar surface, in the lunar module "Snoopy."

[Photo and caption credit: NASA]

Apollo 11

On 20 July 1969, humans from Earth visited another world marking the achievement of a goal set by President John F. Kennedy in 1961 before any Americans had gone into orbit. The landing included avoiding a lunar crater filled with boulders prior to touchdown. Neil Armstrong and Buzz Aldrin explored their landing site for more than two hours, collecting soil and rock samples, setting up experiments, and raising an American flag. They also left behind a plaque on the lunar module's ladder stating, "We came in peace for all mankind."

Above: Portrait of the prime crew of the Apollo 11 lunar landing mission. From left to right: commander Neil A. Armstrong, command module pilot Michael Collins, and lunar module pilot Edwin E. Aldrin Jr. The Apollo 11 crew became the first people to land on the surface of the moon.

[Photo and caption credit: Alamy Stock Photo]

Opposite: July 2009, Washington, D.C. From left to right: former astronauts Michael Collins, Neil Armstrong, and Buzz Aldrin attend a ceremony on Capitol Hill, honoring the Apollo 11 mission in which Armstrong was the first human to step on the moon.

[Photo and caption credit: Alamy Stock Photo]

Neil Armstrong: First Man on the Moon

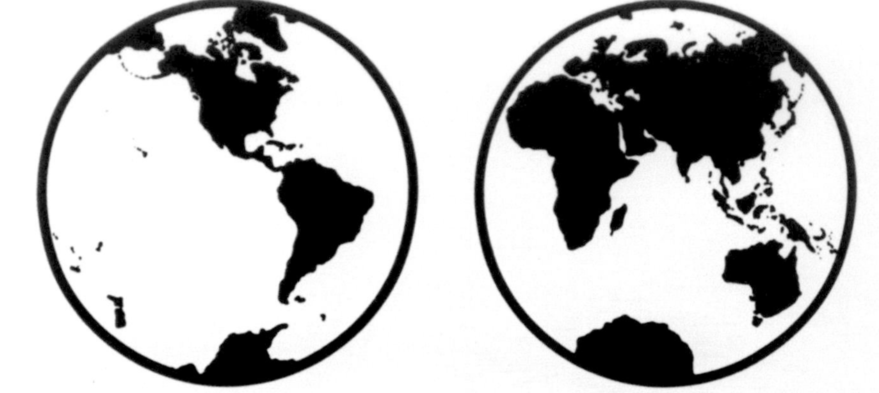

HERE MEN FROM THE PLANET EARTH
FIRST SET FOOT UPON THE MOON
JULY 1969, A. D.
WE CAME IN PEACE FOR ALL MANKIND

NEIL A. ARMSTRONG
ASTRONAUT

MICHAEL COLLINS
ASTRONAUT

EDWIN E. ALDRIN, JR.
ASTRONAUT

RICHARD NIXON
PRESIDENT, UNITED STATES OF AMERICA

Top right: A close-up view of an astronaut's boot print in the lunar soil, photographed with a 70mm lunar surface camera during the Apollo 11 extravehicular activity on the Moon.

Bottom right: This is a replica of the plaque which the Apollo 11 astronauts left behind on the Moon in commemoration of the historic event. The plaque is made of stainless steel and was attached to the ladder on the landing gear strut on the descent stage of the Apollo 11 Lunar Module.

Opposite: Neil A. Armstrong, commander for the Apollo 11 Moon-landing mission, practices for the historic event in a lunar module simulator in the Flight Crew Training Building at Kennedy.

[Photo and caption credits: NASA]

Apollo 12

The encore Apollo lunar landing mission took place from 14–24 November 1969. The plan for Apollo 12 included recovering pieces of Surveyor 3, a robotic lander that landed on the Moon two years earlier. Scientists were able to study the effects of exposure in the lunar environment. After a pinpoint landing that gave the crew easy access to Surveyor, they also deployed an experiments package. One of the instruments was a seismometer, an instrument that responds to ground phenomena such as moonquakes. Before leaving lunar orbit, they jettisoned the lunar module's upper stage so it would crash on the surface, providing a controlled experiment for the seismometers.

Right: A close-up view of lunar sample 12,052 under observation in the Manned Spacecraft Center's Lunar Receiving Laboratory (LRL). Astronauts Charles Conrad Jr. and Alan L. Bean collected several rocks and samples of finer lunar matter during their Apollo 12 lunar landing mission extravehicular activity. This particular sample was picked up during the second spacewalk on 20 November. It is a typically fine-grained crystalline rock with a concentration of holes on the left part of the exposed side. These holes are called vesicles and have been identified as gas bubbles formed during the crystallization of the rock. Several glass-lined pits can be seen on the surface of the rock. [Photo and caption credit: NASA]

Above: The Apollo 12 Lunar Module (LM), in a lunar landing configuration, is photographed in lunar orbit from the command and service modules (CSM) on 19 November. The coordinates of the center of the lunar surface shown in the picture are 4.5 degrees west longitude and 7 degrees south latitude. The largest crater in the foreground is Ptolemaeus; the second largest is Herschel. Aboard the LM were astronauts Conrad and Bean, with astronaut Richard R. Gordon Jr., command module pilot, remaining with the CSM in lunar orbit while Conrad and Bean descended in the LM to explore the surface of the Moon.

[Photo and caption credit: NASA]

Apollo 13

Apollo 13 has been called a "successful failure." Jim Lovell, Jack Swigert, and Fred Haise made it home safely after an explosion crippled their ship resulting in the crew never landing on the Moon. A switch and insulation for an oxygen tank were damaged during testing as part of the manufacturing process. Both should have been modified during an upgrade to the fuel cell tank that helped create electrical power. When the mechanism to periodically stir the cryogenic liquid oxygen was switched on during flight, the tank exploded. This depleted almost all of the electrical power from the command module, forcing the crew to use the lunar module as a "lifeboat." The astronauts made it home safely thanks to the mission control team's perseverance developing and implementing improvised procedures to ensure the astronauts safe return.

Right: President Richard M. Nixon and the Apollo 13 crew salute the US flag during the post-mission ceremonies at Hickam Air Force Base, Hawaii. Earlier, the astronauts John Swigert, Jim Lovell, and Fred W. Haise were presented the Presidential Medal of Freedom by the chief executive.

[Photo and caption credit: NASA]

Above: This view of the severely damaged Apollo 13 service module (SM) was photographed from the lunar module/command module (LM/CM) following SM jettisoning. As seen in this cropped image, enlarged to provide a close-up view of the damaged area, an entire panel on the SM was blown away by the apparent explosion of oxygen tank number two located in Sector 4 of the SM. The damage to the SM caused the Apollo 13 crew members to use the LM as a "lifeboat."

[Photo and caption credit: NASA]

Apollo 14

The Apollo 14 mission targeted the Moon's Fra Mauro highlands, the site planned for exploration by Apollo 13. It also marked the return to space of Alan Shepard, America's first astronaut. The shuttle launched on 31 January 1971; the crew spent more than nine hours outside the lunar module setting up a number of experiments. Mission commander Shepard and lunar module pilot Ed Mitchell explored the lunar surface, pulling a handcart carrying their tools and rock samples.

Right: A close-up view of two components of the Apollo lunar surface experiments package (ALSEP) which the Apollo 14 astronauts deployed on the Moon during their first extravehicular activity. In the center background is the ALSEP's central station; and in the foreground is the mortar package assembly of the ALSEP's active seismic experiment. The modularized equipment transporter can be seen in the background on the right.

Opposite: The huge, 363-feet tall Apollo 14 space vehicle is launched from Pad A, Launch Complex 39, at the Kennedy Space Center in Florida. This view of the lift off was taken by a camera mounted on the mobile launch tower. Aboard the Apollo 14 spacecraft were commander Alan B. Shepard Jr., command module pilot Stuart A. Roosa, and lunar module pilot Edgar D. Mitchell.

[Photo and caption credits: NASA]

"For when I look at the Moon, I do not see a hostile, empty world. I see the radiant body where man has taken his first steps into a frontier that will never end."

David R. Scott, Commander, Apollo 15

Above: An excellent view of the Apollo 14 lunar module (LM) on the Moon, as photographed during the first Apollo 14 extravehicular activity on 5 February. The astronauts have already deployed the US flag. Note the laser ranging retro reflector (LR-3) at the foot of the LM ladder. The LR-3 was deployed later. While astronauts Shepard and Mitchell descend in the LM to explore the Moon, astronaut Roosa remained with the command and service modules in lunar orbit.
[Photo and caption credit: NASA]

Apollo 15

Lifting off on 26 July 1971, Apollo 15 was the first time humans would drive a vehicle on the Moon. The mission carried a lunar roving vehicle and was the first of the Apollo's missions designed for longer stays on the lunar surface. During their 18 hours on the Moon, commander Dave Scott and lunar module pilot Jim Irwin used this "dune buggy" type vehicle to travel more than 17 miles. They set up experiments and collected 170 pounds of rock and soil samples. Before leaving the lunar surface, Scott conducted an experiment to test a theory developed by astronomer Galileo. He believed that, in a vacuum without air resistance, different objects would fall at the same rate. Scott simultaneously dropped a hammer and a feather with both hitting the ground at the same time. Galileo was right.

Right: This photograph of the lunar roving vehicle (LRV) was taken during the Apollo 15 mission. Powered by battery, the lightweight electric car greatly increased the range of mobility and productivity on the scientific traverses for astronauts. It weighed 462 pounds (77 pounds on the Moon) and could carry two suited astronauts, their gear and cameras, and several hundred pounds of bagged samples. The LRV's mobility was quite high. It could climb and descend slopes of about 25 degrees. The LRV was designed and developed by the Marshall Space Flight Center and built by the Boeing Company.

Opposite: A close-up view of the scientific instrument module (SIM) to be flown for the first time on the Apollo 15 lunar landing mission. Mounted in a previously vacant sector of the Apollo service module, the SIM carries specialized cameras and instrumentation for gathering lunar orbit scientific data. SIM equipment includes a laser altimeter for accurate measurement of height above the lunar surface; a large-format panoramic camera for mapping, correlated with a metric camera and the laser altimeter for surface mapping; a gamma ray spectrometer on a 25-foot extendible boom; a mass spectrometer on a 21-foot extendible boom; X-ray and alpha particle spectrometers; and a subsatellite which will be injected into lunar orbit carrying a particle and magnetometer, and the S-Band transponder.

[Photo and caption credits: NASA]

"It's almost impossible to comprehend the sheer scale of our universe and to imagine all the possibilities for life elsewhere. The work being done by NASA helps us to imagine what might be out there and to realize that we are just a small part of a vast and incredible universe."

Howard Berman, former US representative, California

Above: Apollo 15 lunar module pilot Jim Irwin loaded the lunar rover with tools and equipment in preparation for the first lunar spacewalk at the Hadley-Apennine landing site. The lunar module "Falcon" appears on the left in this image. The undeployed laser ranging retro-reflector lies atop Falcon's modular equipment stowage assembly.

[Photo and caption credit: NASA]

Apollo 16

On the Apollo 16 mission, between 16–27 April 1972, mission commander John Young and lunar module pilot Charlie Duke used the lunar roving vehicle during their three moonwalks. While driving more than 16 miles, they collected about 209 pounds of lunar rocks and soil samples. Problems with the command module's engine steering system prior to the landing, forced mission controllers to cut the mission short by a day. Nevertheless, command module pilot Ken Mattingly performed a spacewalk during the return trip, collecting film canisters from the scientific instrument module in the spacecraft's service module.

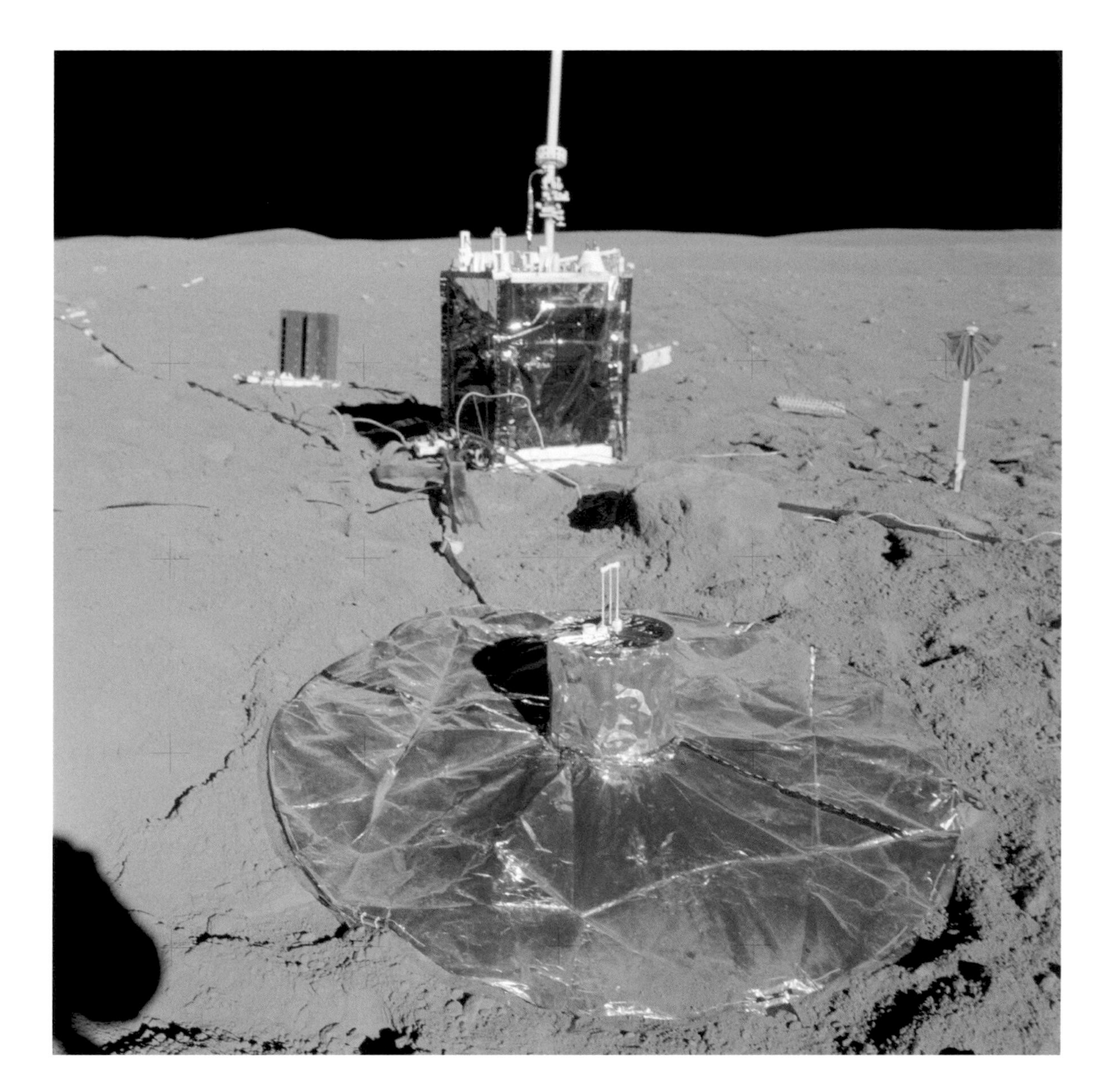

Right: A partial view of the Apollo 16 lunar surface experiments package in deployed configuration on the lunar surface as photographed during the mission's first extravehicular activity, on 21 April 1972. While commander John W. Young and lunar module pilot Charles M. Duke Jr. descended in the Apollo 16 lunar module "Orion" to explore the Descartes highlands landing site on the Moon, command module pilot Thomas K. Mattingly II, remained with the command and service module's "Casper" in lunar orbit.

Opposite: A large crowd of spectators look on as the 363-foot tall Apollo 16 space vehicle moves out of the vehicle assembly building at the Kennedy Space Center's launch complex 39 toward Pad A. The Saturn V stack and its mobile launch tower are atop a huge crawler-transporter.

[Photo and caption credits: NASA]

USA
MOBILE SNACK BAR

Apollo 17

The last Apollo mission to the Moon included the most extensive lunar exploration of the Apollo program. Apollo 17 commander Gene Cernan and lunar module pilot Jack Schmitt stayed on the surface for more than three days, with three moonwalks each lasting more than seven hours. As a geologist, Schmitt was the first scientist–astronaut to go to the Moon. They collected 243 pounds of lunar material. Together with the samples from previous missions, the soil and rocks continue to reveal the Moon's secrets as new tools and techniques are developed.

Right: NASA Apollo 17 crew member Eugene Cernan drives the lunar roving vehicle around the lunar mo—dule during extravehicular activity on the lunar surface on 11 December 1972 at the Taurus-Littrow landing site on the Moon.

Opposite: The prime crew of Apollo 17, photographed with a lunar roving vehicle trainer. They are commander Eugene A. Cernan (seated), command module pilot Ronald E. Evans (standing on right), and lunar module pilot Harrison H. Schmitt (standing on left).

[Photo and caption credits: NASA]

USA

Right: 1972, US astronaut Harrison H. Schmitt standing beside a large lunar boulder on the Moon during the third Apollo 17 extravehicular activity. The crosses that can be seen on this photo (and others in the book) are part of the original image. They allow for correcting film distortion, and help in judging sizes and distances of objects.

[Photo and caption credits: Alamy]

23

"When I circled the
Moon and looked back
at Earth, my outlook on
life and my viewpoint of
Earth changed...Earth is a
spaceship, just like Apollo—
and just like Apollo, the
crew must learn to live
and work together. We
must learn to manage the
resources of this world with
new imagination."

Commander James Lovell, Apollo 13

Above: NASA's Lunar Reconnaissance Orbiter captured this unique view of Earth from the spacecraft's vantage point in orbit around the Moon.

[Photo and caption credit: NASA]

3: The Space Shuttle Program

By the early 1970s, the United States had been launching humans into space for a decade. As NASA's Apollo Program was drawing to a close, a design for a new spacecraft began to emerge, a reusable "space truck" that could be used again and again. It was a vehicle that would launch like a rocket and return to Earth like an aircraft. It was called the Space Shuttle.

I can think of no other plane that ever had such a beloved personality as the Space Shuttle. There was never anything like it and when it launched, with two external fuel tanks and a single massive fuel tank, it was majestic and awe-inspiring. With 135 missions, filled with accomplishments and tragedies, this vehicle set the foundation for deep space exploration, space colonies, and the International Space Station, while also conducting thousands of experiments whose benefits continue to be invaluable.

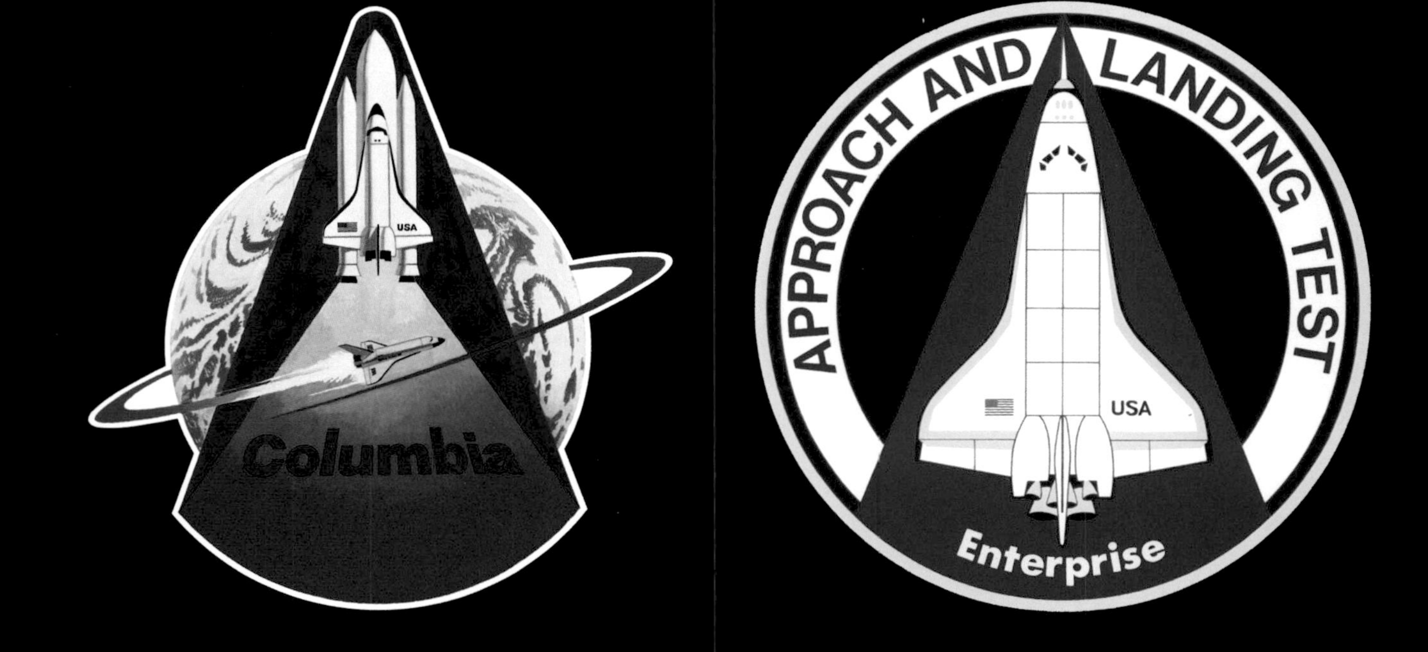

USA
Columbia
APPROACH AND LANDING TEST
USA
Enterprise

COLUMBIA
CHALLENGER
DISCOVERY
ATLANTIS
ENDEAVOUR

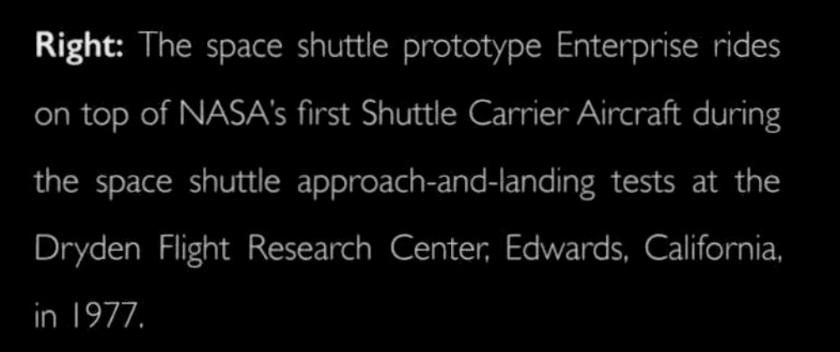

Right: The space shuttle prototype Enterprise rides on top of NASA's first Shuttle Carrier Aircraft during the space shuttle approach-and-landing tests at the Dryden Flight Research Center, Edwards, California, in 1977.

Enterprise
NASA
United States

mounted onto a test version of an external tank and solid rocket boosters, the "stack" was rolled from the Vehicle Assembly Building to the Launch Complex 39A where Apollo Moon landing missions began their flights to the Moon.

After testing to pave the way for the first shuttle to fly in space, Enterprise was returned to Edwards on 6 September 1981 for storage. Enterprise was next used following the landing of STS-4. It was a backdrop as President Ronald Reagan welcomed Ken Mattingly and Hank Hartsfield home on the 4 July 1982. A year later, Enterprise was sent on a tour of several nations in Europe, including the United Kingdom, Germany, Italy, and the Paris Air Show in France. During the return trip, Enterprise stopped in Ottawa, Canada, before returning to NASA's Dryden Research Center at Edwards, on 13 June, again going into storage.

While never flown in space, Enterprise supported its sister ships that did. In 1987, NASA tested an arresting barrier to be used in the event that the space shuttle's brakes failed upon landing. In a test at Dulles International Airport in Virginia, technicians and engineers winched Enterprise onto a landing barrier. Later, in 1987, NASA used the original shuttle to test crew-bailout procedures established following the 1986 loss of Space Shuttle Challenger. Engineers also removed portions of Enterprise's structure during the Space Shuttle Program for examination and study. Evaluations included testing the structural integrity of the payload bay doors. Results of the appraisals determined that the doors were in good condition following years in storage. In April 2003, after the shuttle Columbia and its crew were lost on re-entry, investigators used Enterprise's left landing-gear door and part of a wing to test the results of impacts by foam from the external fuel tank. The trials proved that a foam strike was the cause of the accident.

Above: President Ronald Reagan (in tan suit on podium, waving) acknowledges the cheers of thousands of flag-waving spectators gathered in front of the prototype Space Shuttle Enterprise at the back ramp at NASA's Dryden Flight Research Center, following the landing of Columbia on the fourth shuttle mission, 4 July 1982.

Below: Space Shuttle Enterprise at the Palmdale manufacturing facilities with the cast and crew of the television series, *Star Trek*. From left to right: DeForest Kelley (Dr "Bones" McCoy), George Takei (Mr Sulu), James Doohan ("Scotty"), Nichelle Nichols (Lt. Uhura), Leonard Nimoy (Mr Spock); the series creator Gene Roddenberry; NASA deputy administrator, George Low; and Walter Koenig (Ensign Pavel Checkov).
[Photo and caption credit: NASA]

Opposite: A yellow sling is lowered onto Space Shuttle Enterprise, which sits atop NASA's 747 Shuttle Carrier Aircraft prior to it being demated a few hours later at John F. Kennedy International Airport in New York on Saturday 12 May 2012. The shuttle was placed on a barge and moved by tugboat up the Hudson River where it was then lifted by crane and placed on the flight deck of the Intrepid.
[Photo and caption credit: NASA/Kim Shiflet]

On 20 November 2003, Space Shuttle Enterprise was moved from a hangar at Dulles Airport to a new extension of the Smithsonian National Air and Space Museum, the Stephen F. Udvar-Hazy Center. A month later, experts restored the 26-year-old orbiter going on display. On 8 July 2011, Space Shuttle Atlantis touched down at NASA's Kennedy Space Center with STS-135 being the final mission. The orbiters were retired and delivered to museums. Following post-flight de-servicing of Space Shuttle Discovery, the orbiter was mounted atop the Shuttle Carrier Aircraft and transported to the Udvar-Hazy Center on 17 April 2012. With the arrival of the orbiter, that completed 39 missions and Enterprise was transferred from Dulles Airport to New York City where it is now on display at the Intrepid Sea, Air, and Space Museum. When Enterprise landed at John F. Kennedy International Airport, the cast and crew of the original *Star Trek* television series and motion pictures were there. Leonard Nimoy, who played Mr Spock, had been on hand in 1976 when Enterprise originally rolled out of the Rockwell facility in California.

ATC 007
LOMMA

Space Shuttle Columbia

A new era in human spaceflight began on 12 April 1981, as Space Shuttle Columbia lifted off with astronauts John Young and Bob Crippen aboard. America's first reusable space vehicle was named for an oceangoing ship that circled the world as well as the command module for the Apollo 11 Moon landing. Known as STS-1, for Space Transportation System-1, Columbia orbited the Earth for two days, landing at Edwards Air Force Base, successfully demonstrating that all primary systems performed as designed.

One of the features of NASA's space shuttle fleet was flexibility. When upgraded, the vehicles performed better as new technology became available. During servicing in 1991, Columbia received numerous improvements, including an enhanced thermal protection system for better protection on re-entry. Later, the shuttle's systems were upgraded with a state-of-the-art, multi-functional electronic display system known as a "glass cockpit."

Over the following 22 years, Columbia served as the "flagship" of the space shuttle fleet. The orbiter flew many scientific research missions using a space station precursor research facility called Spacelab.

Opposite: Space Shuttle Columbia begins a new era of space transportation when it lifts off from NASA Kennecy Space Center. The reusable orbiter, with its two fuel tanks and two solid rocket boosters, has just cleared the launch tower. Aboard the spacecraft are commander John W. Young and pilot Robert L. Crippen. [Photo and caption credit: NASA]

Left: Just prior to dawn on 7 December 1996, Space Shuttle Columbia heads for a landing on Runway 33 at the Kennedy Space Center's (KSC) Shuttle Landing Facility to successfully complete a 17-day mission. The landing is the 33rd at KSC for the Space Transportation System. Crew members aboard were mission commander Kenneth D. Cockrell and pilot Kent V. Rominger; along with mission specialists Story Musgrave, Tamara E. Jernigan, and Thomas D. Jones.

Spacelab

The STS-9 space shuttle mission of Columbia launched on 26 November 1983, with the first flight of Spacelab, a reusable laboratory developed by the European Space Agency. Mounted inside the orbiter's payload bay, Spacelab was attached to the crew module giving crewmembers the opportunity to perform research and experiments in a "shirt-sleeve" environment. The workshop supported studies in biology, materials science, astronomy, and other disciplines.

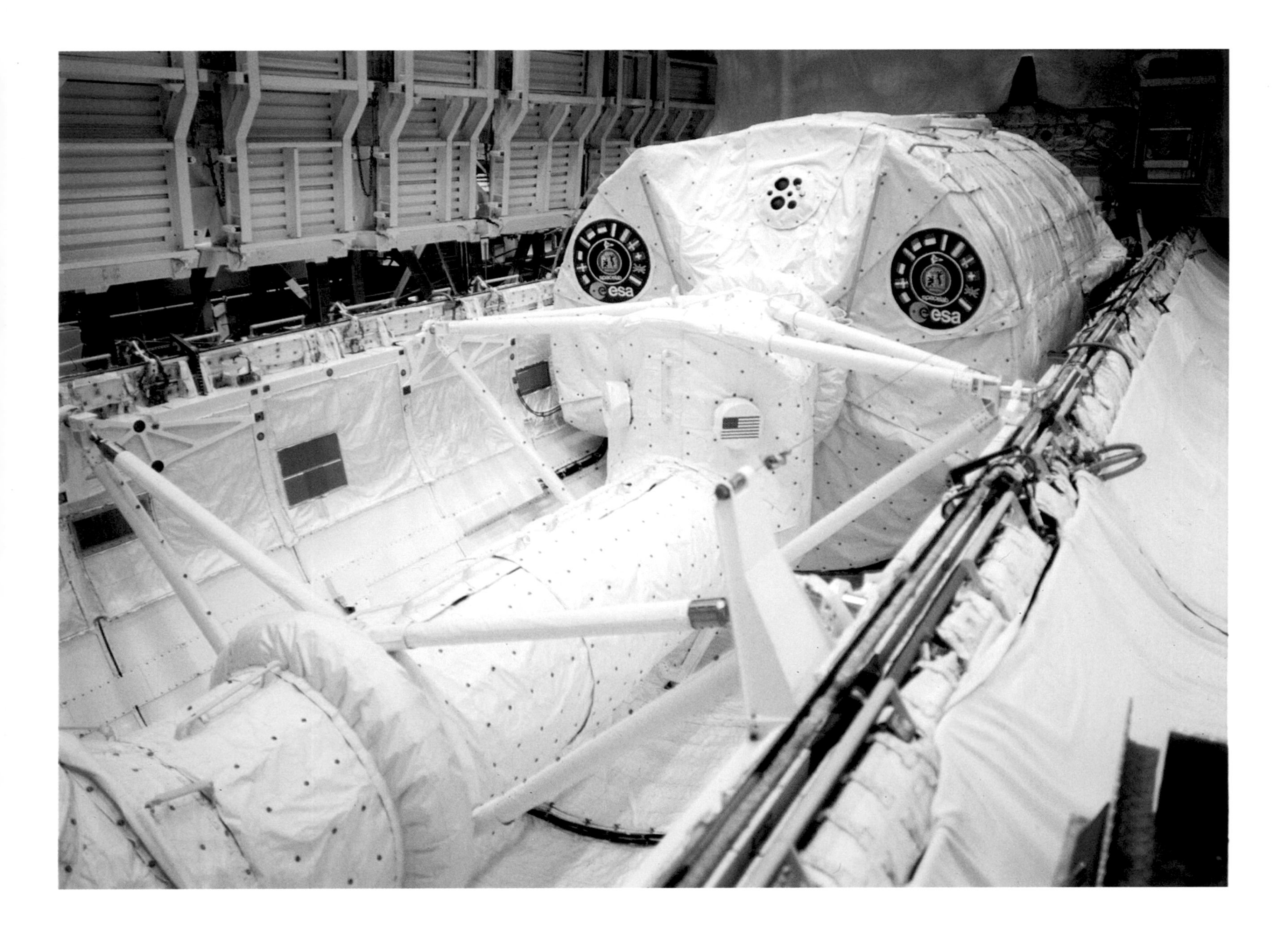

Opposite: This photograph shows the Spacelab-1 module and Spacelab access tunnel being installed in the cargo bay of orbiter Columbia for the STS-9 mission. The orbiting laboratory, built by the European Space Agency, was capable of supporting many types of scientific research best performed in space. The Spacelab access tunnel, the only major piece of Spacelab hardware made in the USA, connected the module with the mid-deck level of the orbiter cabin.

[Photo and caption credit: NASA]

Above: Astronauts Joseph Kerwin (left) and William Lenoir familiarize themselves with equipment aboard the Spacelab mock-up during a 1976 visit to the Marshall Space Flight Center. Kerwin and Lenoir were part of an astronaut group briefed on Spacelab subsystems and crew activities by Marshall scientists and engineers. The Marshall Space Flight Center had management responsibility for Spacelab.

[Photo and caption credit: NASA]

Above right: Ready to begin one of her busy 12-hour shifts, payload specialist Dr. Chiaki Naito-Mukai enters the International Microgravity Laboratory (IML-2) science module in the cargo bay via the tunnel connecting it to Space Shuttle Columbia's cabin. Dr. Muka , representing the National Space Development Agency (NASDA) of Japan, joined six NASA astronauts for more than two weeks of experimenting in Earth orbit.

[Photo and caption credit: NASA]

A Series of Columbia Firsts

In addition to STS-1 and the maiden flight of Spacelab, Columbia firsts include:

• Dr. Ulf Merbold of Germany became the first European Space Agency astronaut in space when he took part in the STS-9 mission.

• On STS-61C, Columbia's crew included U.S. Rep. Bill Nelson (D-Fla.), the first member of Congress to fly in space. Nelson went on to become NASA's administrator.

• The Japanese Space Agency's Chiaki Mukai was the first Japanese woman in space when she went on the STS-65 mission in July 1994.

• During a TV broadcast from space, the STS-73 astronauts aboard Columbia "threw out" the ceremonial first pitch for game five of the 1995 World Series taking place in Cleveland, with the "pitcher" literally outside of the world.

Left: Payload specialist Ulf Merbold shown working in the Spacelab 1 module onboard Space Shuttle Columbia.
[Photo and caption credit: NASA]

Chandra X-ray Observatory

One of the more important Columbia missions was STS-93. NASA astronaut Eileen Collins served as the mission's commander, the first woman to do so. The primary objective for the flight was deployment of the Chandra X-ray Observatory after the crew reached orbit on 23 July 1999. Still in flight today, the advanced X-ray astrophysics facility was designed to observe deep-space objects such as previously invisible black holes and high-temperature gas clouds.

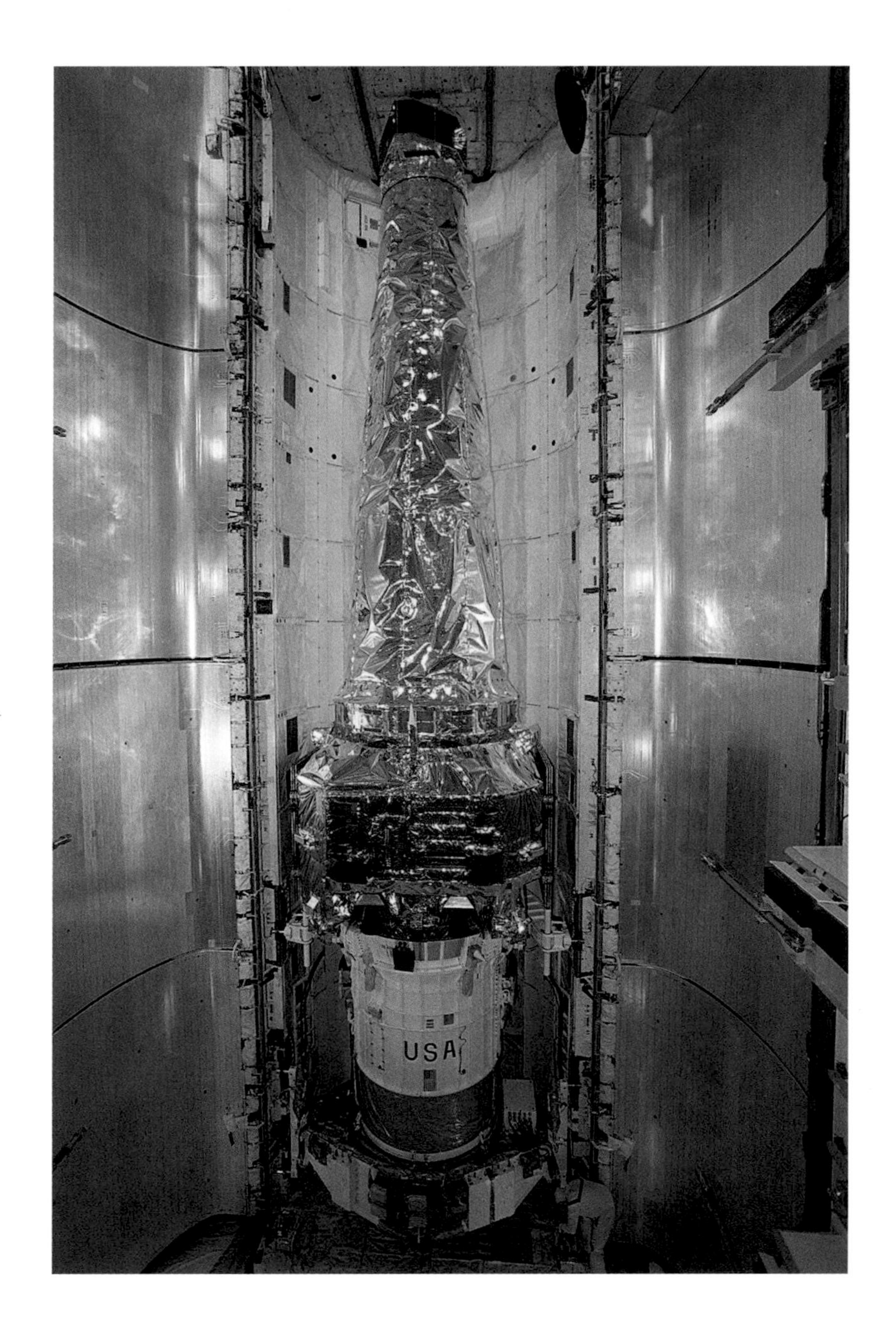

Right: In July 1999, Space Shuttle Columbia delivered the Chandra X-Ray Observatory—shown here being installed and mated to the inertial upper stage inside Columbia's cargo bay at NASA's Kennedy Space Center—to low-Earth orbit.

Opposite: In this fish-eye view, a worker oversees the movement of the Chandra X-ray Observatory into the payload bay of the orbiter Columbia. Chandra was the primary payload on mission STS-93, which took place in July 1999 aboard Space Shuttle Columbia. The world's most powerful X-ray telescope, Chandra allows scientists from around the world to see previously invisible black holes and high-temperature gas clouds, giving the observatory the potential to rewrite the books on the structure and evolution of our universe.

[Photo and caption credits: NASA]

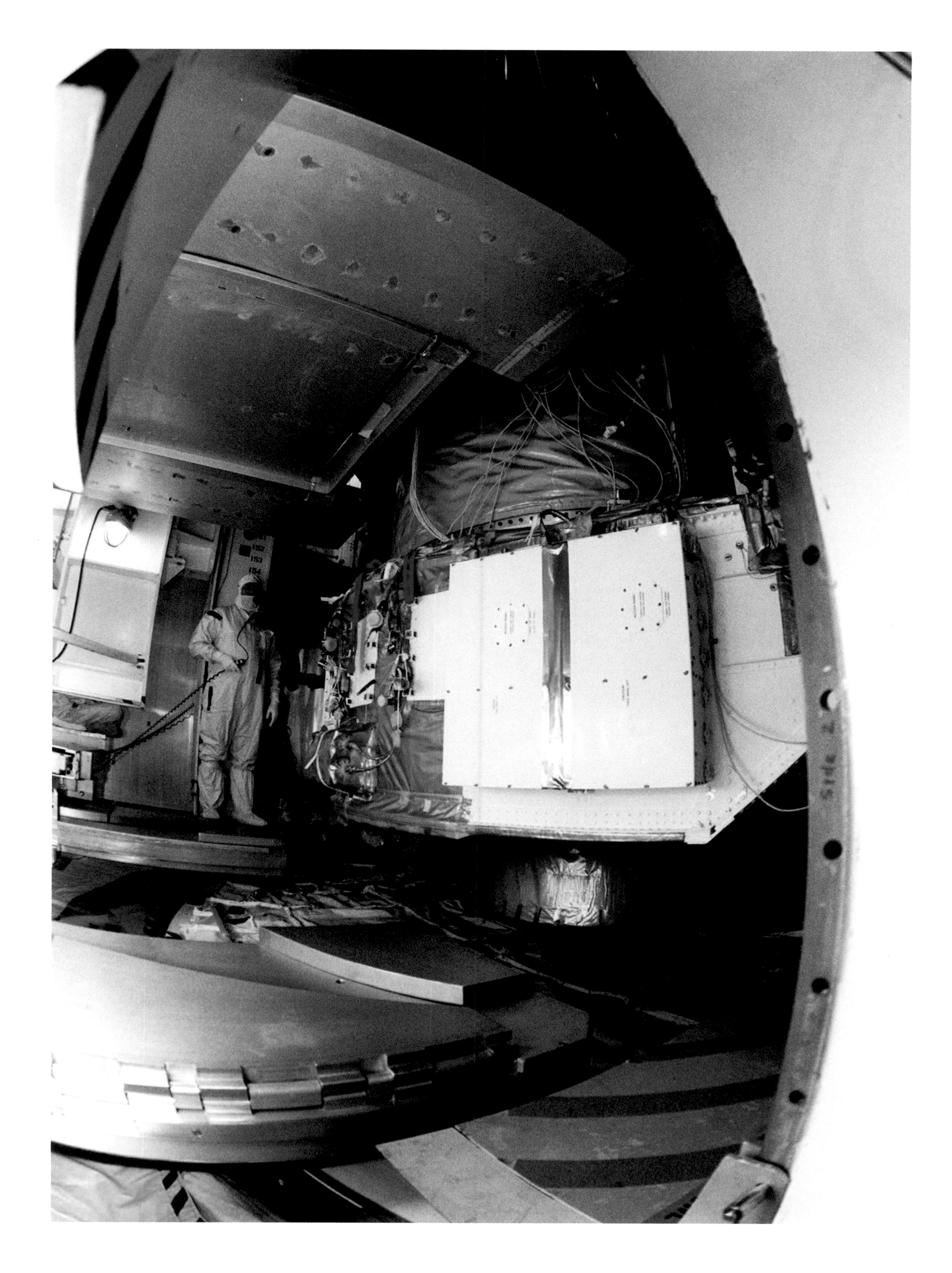

"If you want to find the secrets of the universe, think in terms of energy, frequency, and vibrations."

Nikola Tesla, inventor, electrical and mechanical engineer, and futurist

Above: This photograph shows the mirrors of the High Resolution Mirror Assembly (HRMA) for the Chandra X-Ray Observatory (CXO), formerly Advanced X-Ray Astrophysics Facility (AXAF), being assembled in the Eastman Kodak Company in Rochester, New York. The AXAF was renamed CXO in 1999. The CXO is the the world's most sophisticated and most powerful x-ray telescope ever built. It observes x-rays from high-energy regions of the universe, such as hot gas in the remnants of exploded stars. The HRMA, the heart of the telescope system, is contained in the cylindrical "telescope" portion of the observatory. Since high-energy x-rays would penetrate a normal mirror, special cylindrical mirrors were created. The two sets of four-nested mirrors resemble tubes within tubes. Incoming x-rays graze off the highly polished mirror surface and are funnelled to the instrument section for detection and study.

[Photo and caption credit: NASA]

Mission STS-51L

NASA was planning for 1986 to be their most ambitious year to date, launching as many as 15 missions, including deployment of the Hubble Space Telescope and the first polar-orbiting flight launched from Vandenberg Air Force Base in California.

After several delays, STS-61C launched on 12 January 1986. Next up was STS-51L, a six-day mission to deploy the second Tracking and Data Relay System, or TDRS-B, a large communications satellite. Additionally, the crew would deploy and retrieve an astronomy payload to study Halley's Comet. The mission was also to include the first "civilian" launched into space. A high school teacher, Christa McAuliffe, was to conduct lessons for children from orbit.

Above: Just moments following ignition, Space Shuttle Challenger, mated to its two solid rocket boosters and an external fuel tank, soars toward a week-long mission in Earth orbit. Note the diamond-shock effect in the vicinity of the three main engines. Launch occurred at 5.00 pm (EDT) on 29 July 1985.

Right: This orbiter tribute of Space Shuttle Discovery, or OV-103, hangs in Firing Room 4 of the Launch Control Center at NASA's Kennedy Space Center in Florida. Discovery's accomplishments include the first female shuttle pilot, Eileen Collins, on STS-63, John Glenn's legendary return to space on STS-95, and the celebration of the 100th shuttle mission with STS-92. In addition, Discovery supported a number of Department of Defense programs, satellite deploy-and-repair missions and 13 International Space Station construction and operation flights. The tribute features Discovery demonstrating the rendezvous pitch maneuver on approach to the International Space Station during STS-114. Having accumulated the most space shuttle flights, Discovery's 39 mission patches are shown circling the spacecraft. The background image was taken from the Hubble Space Telescope, which launched aboard Discovery on STS-31 and serviced by Discovery on STS-82 and STS-103. The American Flag and Bald Eagle represent Discovery's two return-to-flight missions—STS-26 and STS-114—and symbolize Discovery's role in returning American astronauts to space.

[Picture and caption credit: NASA/Amy Lombardo]

LINDSEY BOE BARRATT STOTT
STS-133
HARTSFIELD COATS
MULLANE HAWLEY RESNIK WALKER
ALLEN FISHER GARDNER
HAUCK WALKER
LOUNGE HILMERS NELSON
HAUCK COVEY
ENGLE COVEY LOUNGE FISHER VAN HOFTEN
LOUNGE HILMERS NELSON
HAUCK COVEY
COLLINS KELLY
NOGUCHI ROBINSON THOMAS
LAWRENCE CAMARDA
COATS BLAHA
BAGIAN SPRINGER BUCHLI
SHRIVER BOLDEN
HAWLEY McCANDLESS SULLIVAN
RICHARDS CABANA
MELNICK SHEPHERD AKERS
REIGHTLER BUCHLI GEMAR BROWN
CREIGHTON
HILMERS THAGARD
READDY GRABE OSWALD
WALKER CABANA
BLUFORD CLIFFORD
VOSS
CAMERON OSWALD
FOALE COCKRELL OCHOA
READDY NEWMAN
BURSCH
BOLDEN SEGA REIGHTLER
DAVIS KRIKALEV
RICHARDS HAMMOND
LEE MEADE
LINENGER HELMS
HENRICKS
KREGEL THOMAS
CURRIE WEBER
STS-70

Right: Docked already with Russia's Mir Space Station and backdropped against a half globe of Earth featuring the Crimean Peninsula, Space Shuttle Atlantis is partially visible through a window on the Kvant 2 Module. The crew cabin and forward cargo bay of Atlantis are most prominent. Below center can be seen the Androgynous Peripheral Docking System (APDS) and the Kristall module on Mir. The APDS is connected to a port in a tunnel leading to the Spacelab science module in Atlantis' cargo bay. The linkup enabled the seven STS-71 crew members to visit Mir and the three Mir-18 crew members access to Spacelab. The Black Sea lies directly beneath Atlantis, with Ukraine's diamond-shaped Crimean Peninsula immediately to the right of the cockpit. The wide lower course of the Dnieper River can be seen entering the Black Sea at far right. The coast of Romania and Bulgaria lies at a point where the cloud begins at top right. The peninsula of Asia Minor lies across the left of the view, mostly under cloud cover. The Mediterranean Sea is the cloud-free, blue mass beyond. Still further, at about 1,300 miles distance, the north edge of Africa is stretched out as a line across the horizon with its characteristic sandy color. The nose of Atlantis points southwest toward the only outlet of the Black Sea known as the Bosporus.

[Photo and caption credit: NASA]

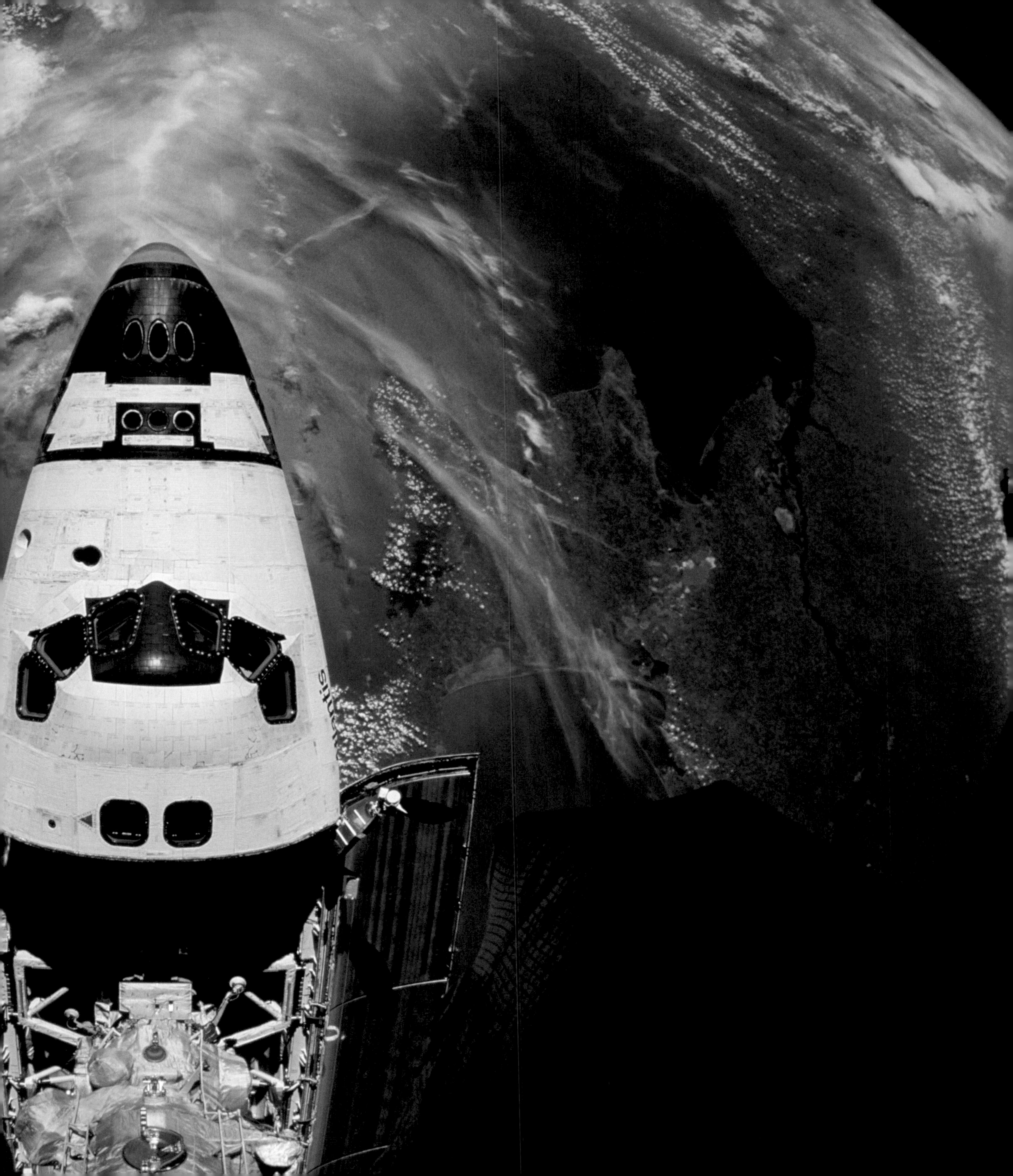

Atlantis–Mir Mission

Atlantis made seven straight flights to Mir, Russia's space station, beginning with STS-71 in 1995. After docking with the orbiting Russian laboratory, Atlantis and Mir became the largest spacecraft in orbit at the time. The mission to Mir included the first on-orbit American crew exchanges, a practice that became common on the International Space Station. Atlantis returned astronaut Shannon Lucid home after her record-breaking 188 days aboard Mir.

Once assembly of the International Space Station began, Atlantis delivered some of the most important elements of the orbiting outpost, including the US laboratory module "Destiny," the Joint Quest Airlock and sections of the integrated truss that serves as the "backbone" of the space station.

A Series of Atlantis Firsts

• On 5 April 1991, STS-37 astronauts Jerry Ross and Jay Apt performed the Space Shuttle Program's first unscheduled spacewalk to assist with a stuck data antenna on the Compton Gamma Ray Observatory.
• STS-71 in 1995 was the first mission in which a shuttle docked with the Russian space station Mir.
• That same mission signaled the 100th American human space flight.
• STS-71 also marked the first on-orbit change-out of shuttle-Mir crewmembers.
• Astronaut Mike Massimino became the first person to use Twitter in space on STS-125 in May 2009.

Above: NASA and the Russian space agency kicked off a new era in international space cooperation in June of 1995, when Space Shuttle Atlantis docked with the Russian space station Mir for the first time. Atlantis' mission, STS-71, launched on 27 June and marked the 100th US human space launch. The Shuttle–Mir program included 11 space shuttle flights and seven astronaut residencies on Mir, and helped pave the way for the International Space Station now in orbit.

[Photo and caption credit: NASA]

"The sky has never been the limit, the sky is only a gateway to the Universe."

Adrin Burruss, friend of the author

Above: It was "installation day" on the International Space Station. The Atlantis and Expedition 13 crews worked on attaching the P3/P4 truss during the first of three scheduled spacewalks by STS-115 shuttle crew members on 12 September 2006. Mission specialist, astronaut Joseph R. Tanner, pictured as he translated along the station hardware, was later joined by mission specialist Heidemarie M. Stefanyshyn-Piper.

[Photo and caption credit: NASA]

Space Shuttle Endeavour

NASA's Space Shuttle Endeavour was built to replace Challenger when it was lost during STS-51L in 1986. It was named after a ship British explorer Captain James Cook sailed to explore the South Pacific in 1768. Following assembly at Rockwell, the new orbiter arrived at the Florida spaceport on 7 May 1991 and launched one year later to the day on its maiden flight.

Endeavour's first mission, STS-49, launched with the ambitious goal to capture and replace the rocket motor on the INTELSAT VI. It was a non-functioning communications satellite needing a boost to reach the necessary higher orbit.

Right: On 7 May 1991, the newest addition to the space shuttle orbiter fleet, Endeavour, arrived at KSC atop the 747 Shuttle Carrier Aircraft. It was demated from the Boeing aircraft and towed to the Vehicle Assembly Building for installation of several major flight components. It would then have a lengthy stay in the Orbiter Processing Facility for a rigorous series of first flow tests.

[Photo and caption credit: NASA]

Above: Exactly one year later, on 7 May 1992, Space Shuttle Endeavour launched into space on its first mission STS-49 from NASA's Kennedy Space Center. The primary mission objective was to capture, repair, and redeploy the Intelsat VI (F-3) communications satellite.

[Photo and caption credit: NASA]

Three-man Spacewalk

In an unprecedented three-person spacewalk by astronauts Rick Hieb, Tom Akers, and Pierre Thuot, they physically grasped the bottom of the large satellite with their gloved hands. As Akers and Hieb took hold, Thuot attached a capture bar so that Bruce Melnick, working inside the shuttle, could use the shuttle's robotic arm to place the Intelsat on a payload kick motor. Later, the new motor propelled the satellite to the correct orbit.

A Series of Endeavour Firsts

• On STS-49 in May 1992, the first three-person spacewalk took place.

• Mae Jemison became the first African American woman in space during STS-47 in September 1992.

• The first Hubble Space Telescope servicing mission was flown during STS-61 in December 1993.

• STS-88 was the first assembly mission for the International Space Station in December 1998.

• First school lessons from orbit by Barbara Morgan took place during STS-118 in August 2007

Opposite: On 13 May 1992, following the successful capture of the Intelsat VI satellite, three astronauts continue moving the 4.5 ton communications satellite into Space Shuttle Endeavour's cargo bay. Left to right, astronauts Richard J. Hieb, Thomas D. Akers and Pierre J. Thuot, cooperate on the effort to attach a specially designed grapple bar underneath the satellite. Thuot stands on the end of the remote manipulator system's arm while Hieb and Akers are on portable foot restraints affixed to Endeavour's portside and the multipurpose support structure, respectively.

[Photo and caption credit: NASA]

NASA
HUGHES
MCDONNELL
DOUGLAS

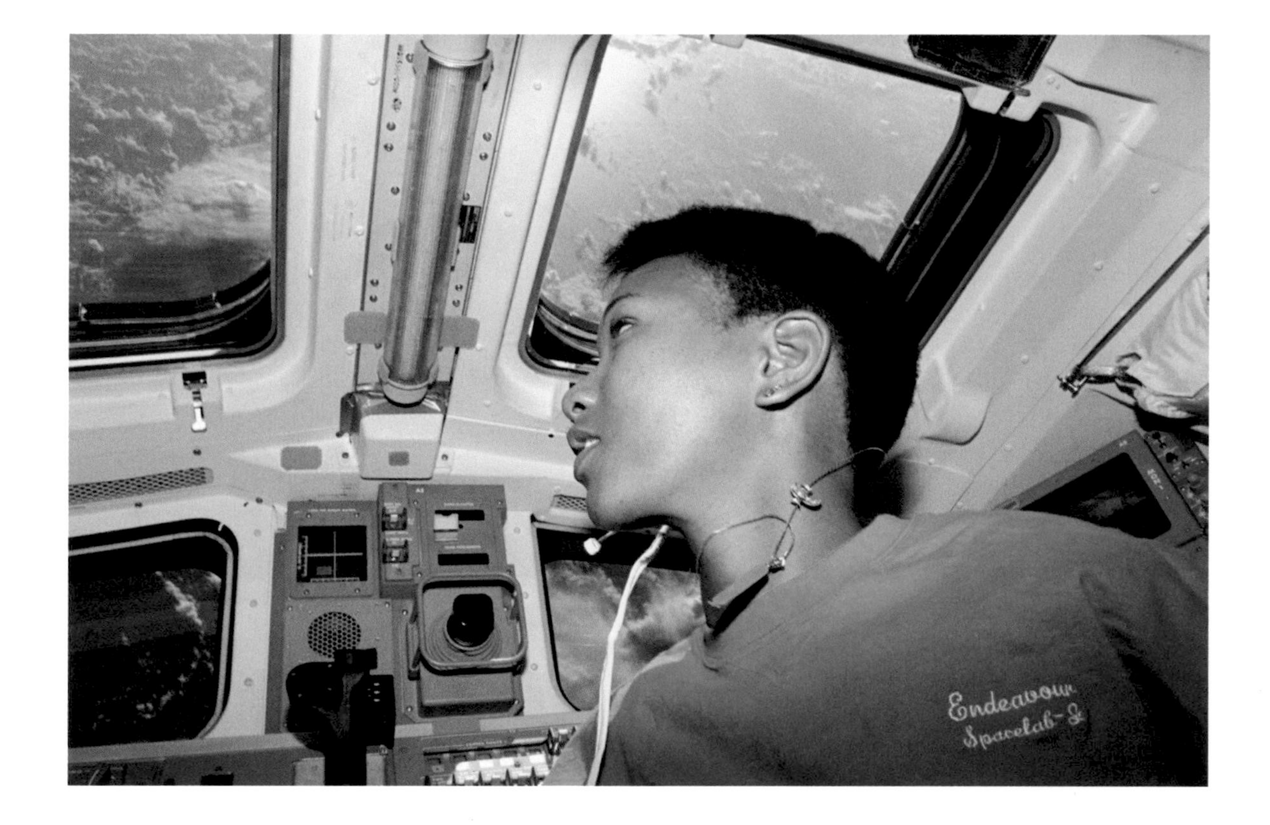

Mae Jemison

Mae Jemison is an engineer, physician, and NASA astronaut who was the first African American woman in space. She was selected as an astronaut in June 1987. On her first and only flight, she was the science mission specialist on STS-47 Spacelab-J. A cooperative effort between the US and Japan, the flight included 44 life-science and materials-processing experiments. Jemison was a co-investigator on a bone-cell research experiment flown on the mission.

Above: In this image, Jemison looks out the aft flight deck ports on Space Shuttle Endeavour's STS-47 mission in 1992.

[Photo and caption credit: NASA]

Above: On 12 September 1992, launch day of the STS-47 Spacelab-J mission on Space Shuttle Endeavour, NASA astronaut Mae Jemison waits as her suit technician, Sharcn McDougle, performs an unpressurized and pressurized leak check on her spacesuit at the Operations and Checkout Building at Kennedy Space Center. Dr. Jemison was the science mission specialist on the eight-day joint mission with Japan's space agency, which included 24 material-science and 20 life-science experiments.

[Photo and caption credit: NASA]

Decommissioning of the Space Shuttle Fleet

Space Shuttle Discovery's final mission concluded with a landing at the Kennedy Space Center on 9 March 2011 after STS-33. Following post-flight de-servicing, the orbiter was mounted atop the Shuttle Carrier Aircraft and transported to the Smithsonian Institution's Steven F. Udvar-Hazy Center in Virginia on 17 April 2012.

The orbiter Enterprise had been a centerpiece display at the Udvar-Hazy Center since 20 November 2003. But with the arrival of Discovery, Enterprise was transported from Dulles Airport to New York City where it is now on display at the Intrepid Sea, Air and Space Museum.

Endeavour concluded its service when it landed at the Florida spaceport on 16 May 2011 at the end of STS-134. After being transported to Los Angeles, Endeavour is now on display at the California Science Center. Plans call for Endeavour to be attached to an external tank and solid rocket boosters and raised in an upright position as if being prepared for launch.

The 30-year Space Shuttle Program ended on 8 July 2011 when Atlantis touched down at Kennedy, with STS-135 being the final mission. Following several months of preparation, the orbiter was towed to the nearby Kennedy Space Center Visitor Complex. It is now on display with cargo bay doors opened and mounted at an angle, giving it the appearance of being in space.

Above: Space Shuttle Discovery on display at the Steven F. Udvar-Hazy Center in Virginia, USA.

[Photo and caption credit: NASA]

4: Skylab to the International Space Station

As scientists began making plans for human exploration of space in the 1950s, an idea emerged for a base in Earth orbit. The concept soon came to be known as a space station. During the late 1960s, NASA managers began developing a program to use Apollo era hardware and technology to build an orbiting laboratory known as Skylab. Lessons learned from that project laid the groundwork for the International Space Station, and so, a multinational station in space was created. If we stop for a moment and think about this accomplishment, while considering the risks and isolation, it's an absolute marvel. Enabling our astronauts to be self-sufficient absolutely sets the stage for permanent space colonies on the Moon—and Mars. The first "mini" International Space Station—Skylab—was launched in 1973 but was only occupied for 24 weeks. The massive work-in-progress International Space Station launched its first segment in 1998, with home-away-from-home facilities providing humans with the ability to become an interplanetary species. Without understanding the mindset for the creation and support of these two stations, we're missing the building blocks of space exploration itself.

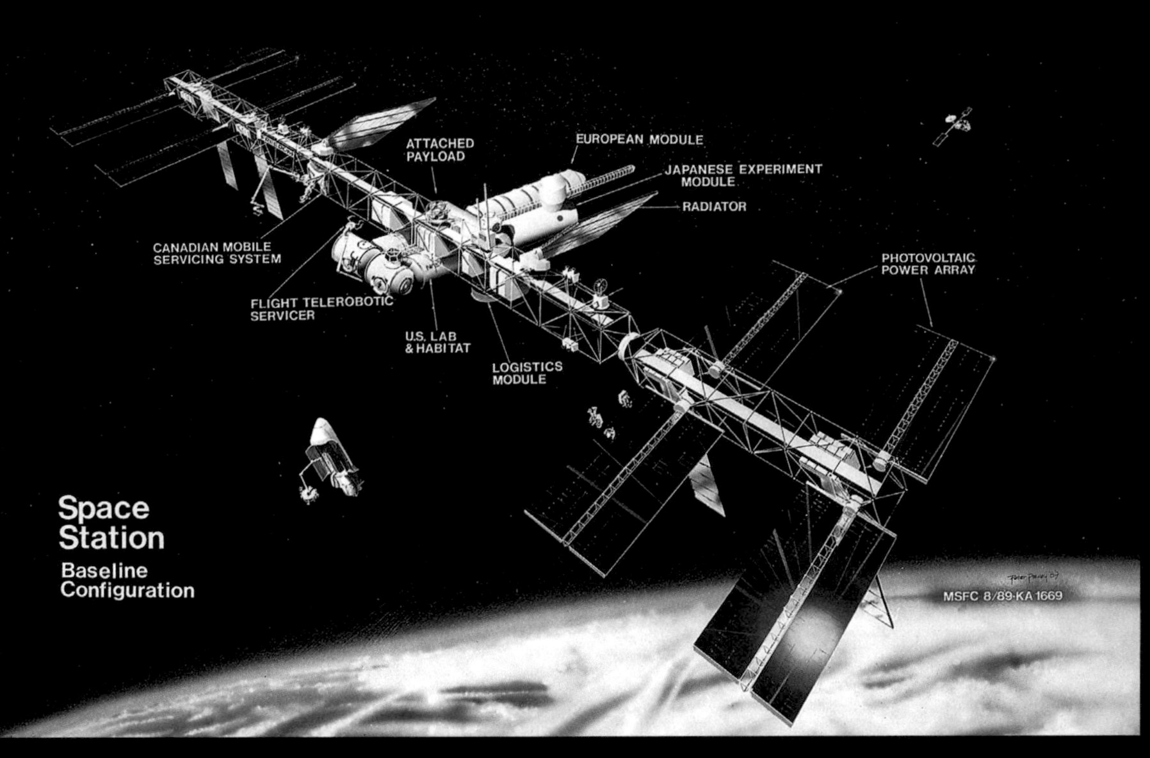

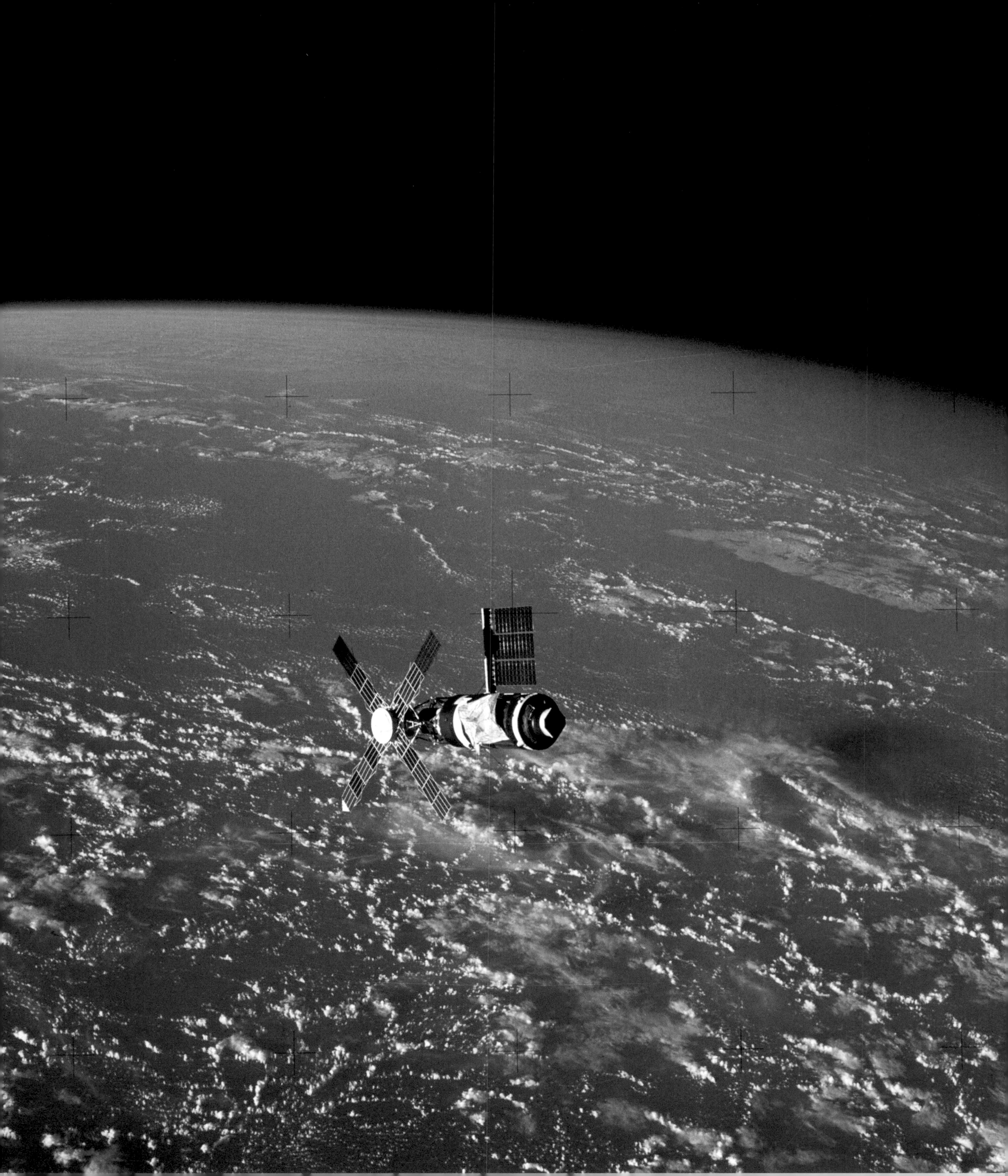

"Experience and knowledge gained from earlier space programs provided a solid foundation on which to build, but the Skylab Program was truly making new pathways in the sky."

Rocco Petrone, director of the Marshall Space Flight Center in the 1970s

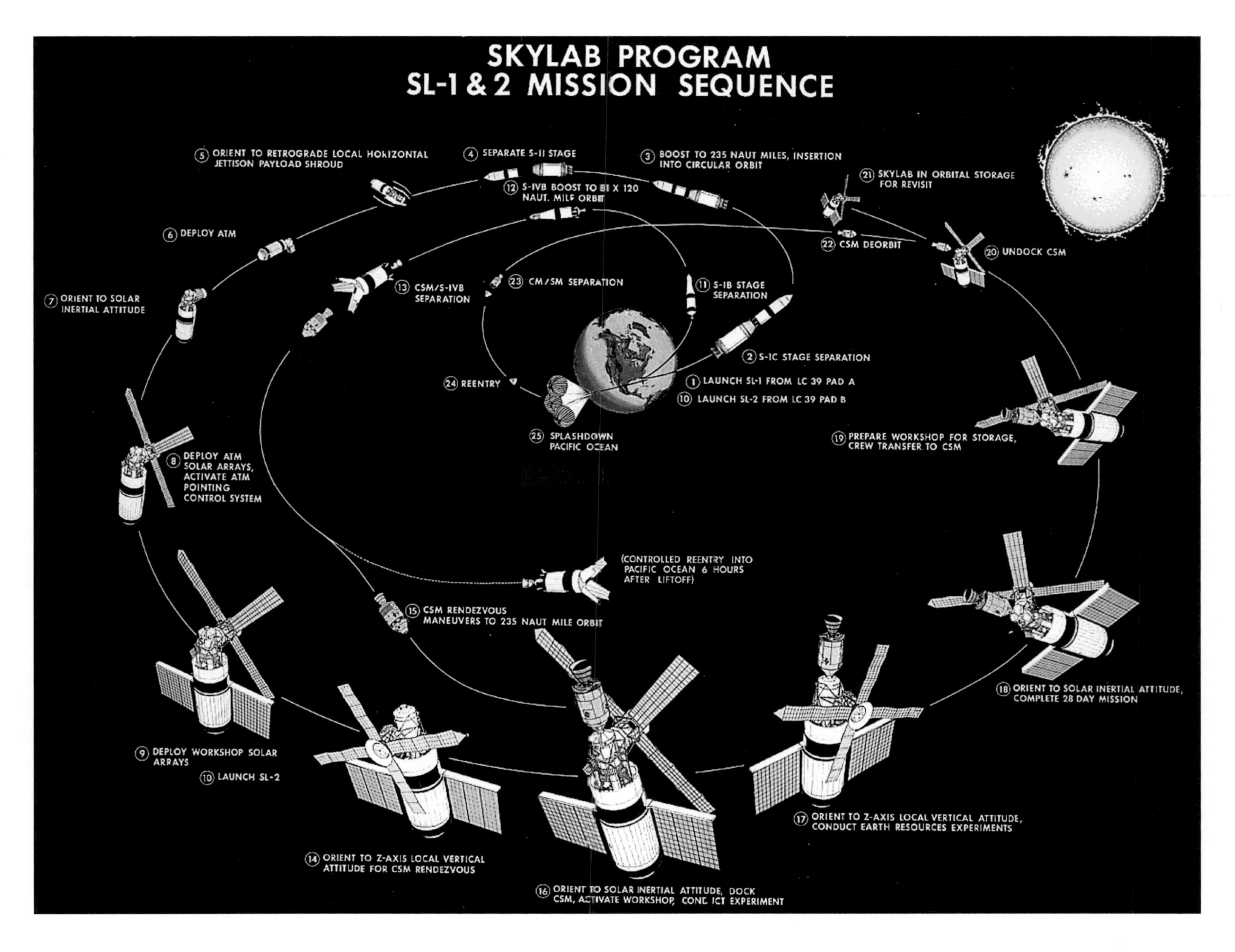

Above: This illustration depicts the Skylab-1 and Skylab-2 mission sequence.

Right: On 2 June 1973, NASA's Marshall Space Flight Center in Huntsville, Alabama, conducted underwater simulations to evaluate possible techniques for freeing jammed solar array panels on the Skylab orbital workshop. A metal strap became tangled over one of the folded solar array panels when Skylab lost its micrometeoroid shield during the launch. Extensive testing and many hours of practice in Marshall's Neutral Buoyancy Simulator helped prepare the Skylab crew for emergency repair procedures in the weightless environment. This photograph shows astronaut Russell Schweickart trying out a bone saw.

[Photo and caption credit: NASA/MSFC]

Apollo Applications Program

The Apollo Applications Program began in 1965 with the aim of developing science-based human space missions using hardware originally developed for the effort to land astronauts on the Moon.

Opposite: This artist's concept is a cutaway illustration of the Skylab with the command/service module being docked to the multiple docking adapter. In an early effort to extend the use of Apollo for further applications, NASA established the Apollo Applications Program (AAP) in August of 1965. The AAP was to include long duration Earth orbital missions during which astronauts would carry out scientific, technological, and engineering experiments in space by utilizing modified Saturn launch vehicles and the Apollo spacecraft.

[Picture and caption credit: NASA]

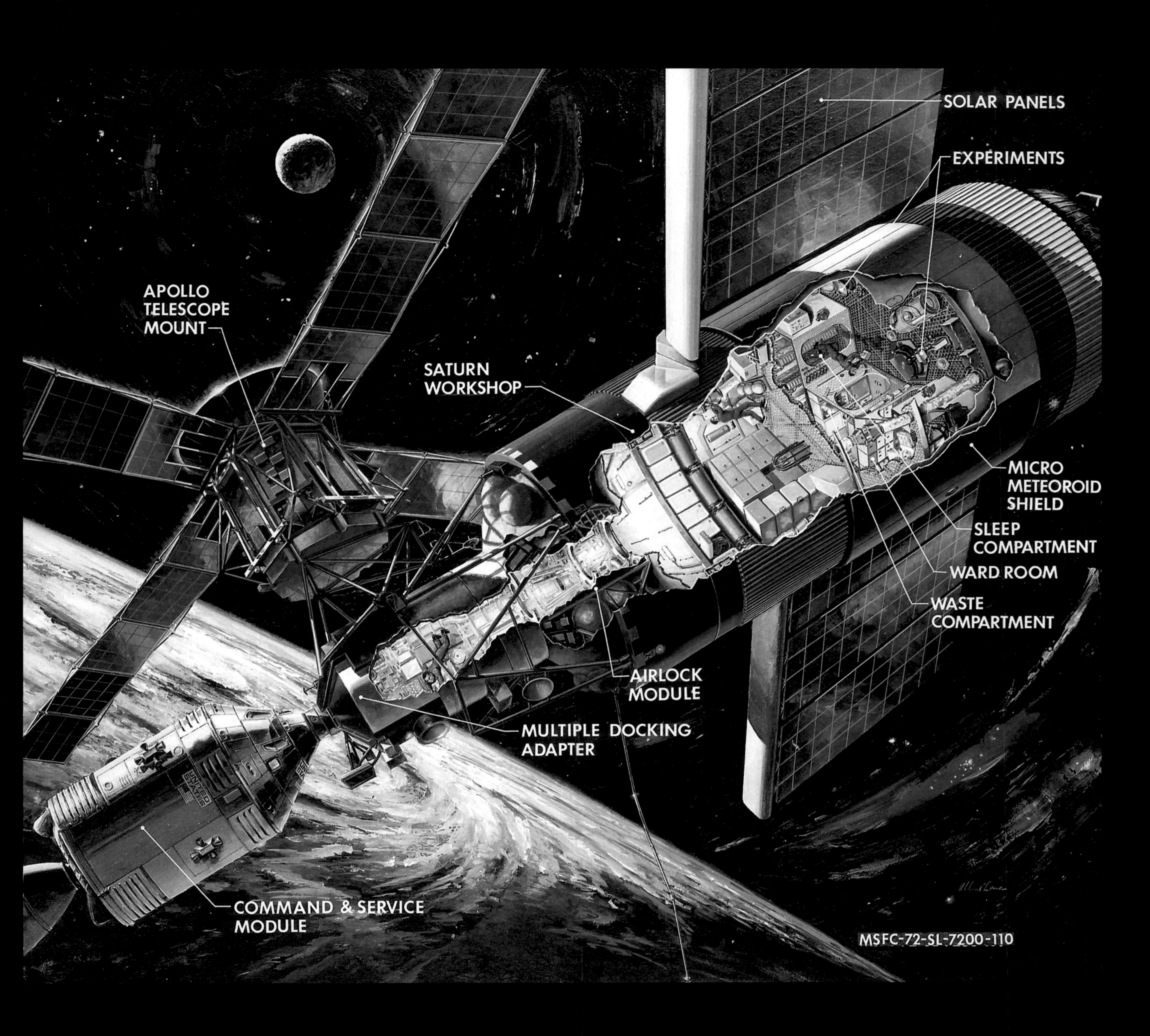
SOLAR PANELS
EXPERIMENTS
APOLLO
TELESCOPE
MOUNT
SATURN
WORKSHOP
MICRO
METEOROID
SHIELD
SLEEP
COMPARTMENT
WARD ROOM
WASTE
COMPARTMENT
AIRLOCK
MODULE
MULTIPLE DOCKING
ADAPTER
COMMAND & SERVICE
MODULE
MSFC-72-SL-7200-110

Skylab 1

As the program matured, its name was changed to Skylab. Engineers at Marshall developed components of the 169,950-pound space station, including an orbital workshop, an airlock module, a multiple docking adapter, an Apollo telescope mount, and systems to allow crews to spend up to 84 days in space. A Saturn V launched the uncrewed Skylab space station, with crews transported atop the smaller Saturn 1B.

Liftoff of the uncrewed Skylab space station took place from NASA's Kennedy Space Center on 14 May 1973. However, it was soon apparent that serious problems had occurred during the launch phase. According to Skylab program manager William Schneider, there was an indication of premature deployment of the meteoroid protective shield. "If that has happened, the shield was probably torn off," he said. "The thermal indications are that it is gone, and we have some indication that our solar array on the workshop also did not fully deploy."

Right: This photograph shows the launch of the SA-513, a modified unmanned two-stage Saturn V vehicle for the Skylab-1 mission, which placed the Skylab cluster into Earth orbit on 14 May 1973. The initial step in the Skylab mission was the launch of a two-stage Saturn V booster, consisting of the S-IC first stage and the S-II second stage, from Launch Complex 39A at the Kennedy Space Center in Florida. Its payload was the unmanned Skylab, which consisted of the orbital workshop, the airlock module, the multiple docking adapter, the Apollo telescope mount, and an instrument unit.

Opposite: An artist's concept illustrating a cutaway view of the Skylab 1 Orbital Workshop (OWS). The OWS is one of the five major components of the Skylab 1 space station cluster which was launched by a Saturn V on 14 May 1973 into Earth orbit.

[Photo and caption credits: NASA]

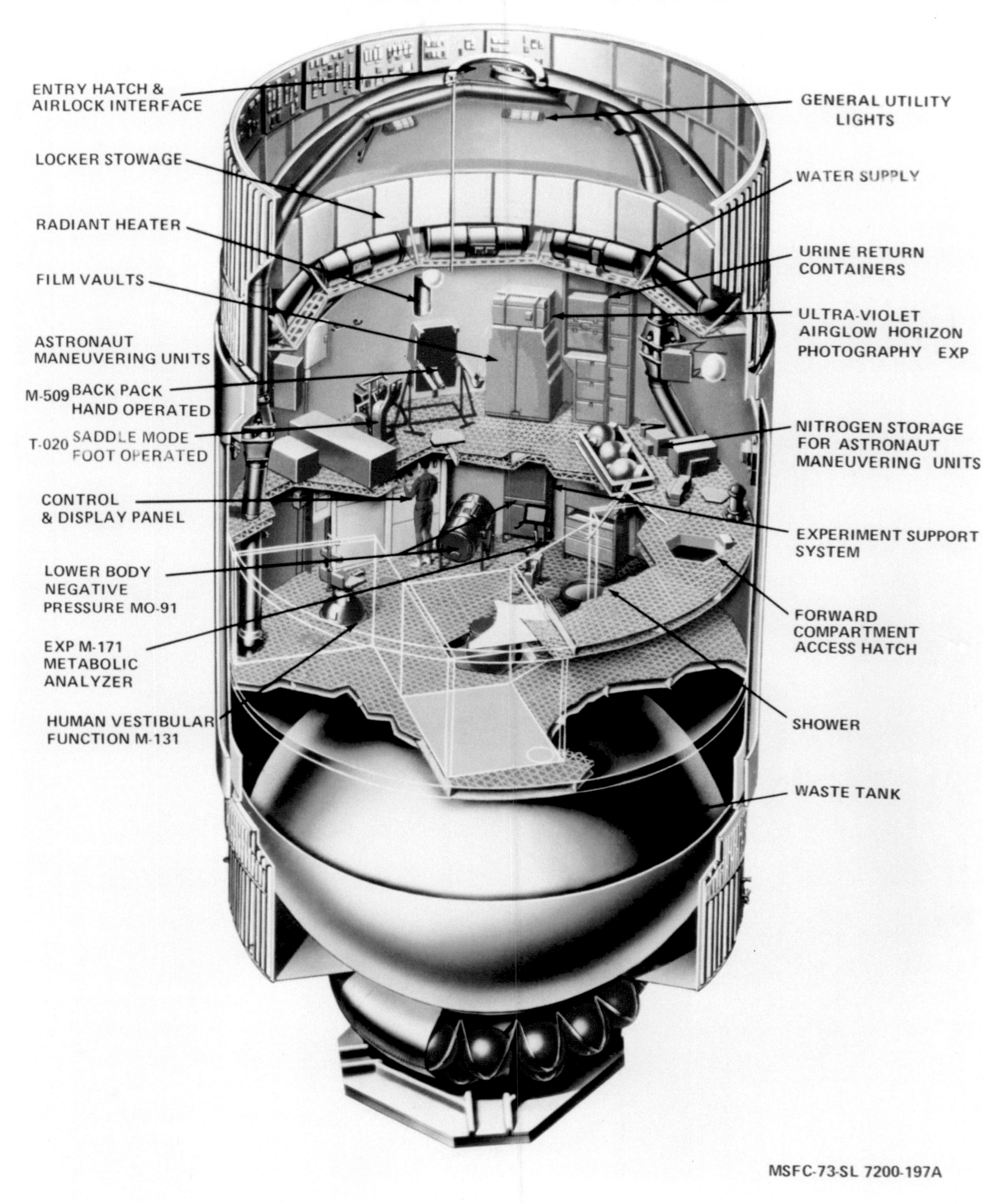
SKYLAB ORBITAL WORKSHOP
ENTRY HATCH &
AIRLOCK INTERFACE
LOCKER STOWAGE
RADIANT HEATER
FILM VAULTS
ASTRONAUT
MANEUVERING UNITS
M-509 BACK PACK
HAND OPERATED
T-020 SADDLE MODE
FOOT OPERATED
CONTROL
& DISPLAY PANEL
LOWER BODY
NEGATIVE
PRESSURE MO-91
EXP M-171
METABOLIC
ANALYZER
HUMAN VESTIBULAR
FUNCTION M-131
GENERAL UTILITY
LIGHTS
WATER SUPPLY
URINE RETURN
CONTAINERS
ULTRA-VIOLET
AIRGLOW HORIZON
PHOTOGRAPHY EXP
NITROGEN STORAGE
FOR ASTRONAUT
MANEUVERING UNITS
EXPERIMENT SUPPORT
SYSTEM
FORWARD
COMPARTMENT
ACCESS HATCH
SHOWER
WASTE TANK
MSFC-73-SL 7200-197A

Skylab 2

Launch of Skylab 2 with the first crew was postponed. NASA and contractors worked feverishly to develop solutions. A square "umbrella-like" thermal shield was developed to shade Skylab from the heat of the Sun. The Skylab 2 crew—Pete Conrad, Joe Kerwin, and Paul Weitz—trained for a spacewalk to free a solar array that was jammed. This would allow it to provide Skylab with required electrical power.

Conrad, Kerwin, and Weitz lifted off on 25 May 1973, soon gaining their first look at the crippled space station. Once inside, they extended the new solar shield through a small scientific experiment airlock on the side normally facing the Sun. The temperatures inside soon lowered to near-normal levels.

During the spacewalk to free the jammed solar panel, Kerwin cut the metal that jammed the solar wing. With a rope sling, Conrad forced the array to deploy. Later, the solar panel fully extended providing electrical power for the remainder of Skylab's missions.

Right: Two members of the prime crew of the first manned Skylab mission assist each other in suiting up in Building 5 at NASA's Johnson Space Center in Houston during a pre-launch training activity. They are scientist and pilot Joseph P. Kerwin (left) and pilot Paul J. Weitz. The third member of the crew was commander Charles "Pete" Conrad Jr.

Opposite: This view of Skylab in orbit was taken by the Skylab 4 (the last Skylab mission) crew.

[Photo and caption credits: NASA]

Mission Accomplished

With repairs in place, the Skylab 2 crew went to work on research in areas such as how different materials can be processed in microgravity, observations of Earth from the vantage point of space, solar astronomy and confirming that humans can operate in space for extended periods. Before Skylab 2, the longest American spaceflight was 14 days by Gemini VII astronauts Frank Borman and Jim Lovell. Skylab 2 doubled that.

After Skylab 2 splashed down in the Pacific Ocean on 22 June 1973 and was recovered, NASA administrator James Fletcher praised the crew and the NASA-industry team: "For the first time, a crew of astronauts has returned from an extended tour in a space laboratory," he said. "Essentially all of the objectives for this mission have been completed."

Skylab 3

Alan Bean, Owen Garriott, and Jack Lousma lifted off for the second crewed mission aboard Skylab 3, launched on 28 July 1973. Garriott and Lousma improved Skylab's solar shading during a spacewalk to install a twin-pole solar shield that provided better thermal control. They returned on 25 September 1973 after spending 59 days in orbit. After the mission, Lousma explained how Skylab spacewalks paid dividends for the future: "We developed the procedures and techniques for doing effective spacewalks on Skylab that were used so successfully in putting together the International Space Station," he said.

Opposite: Here, astronaut Jack R. Lousma participates in extravehicular activity during which he and astronaut Owen K. Garriott deployed a twin pole solar shield, developed by NASA's Marshall Space Flight Center. The shield was needed after the original panel to protect the orbital workshop was ripped off during launch in May 1973. The solution was delivered to the space station just over two months after the first launch, with much of the development and testing performed at Marshall.

[Photo and caption credit: NASA]

LOUSMA
013

Skylab 4

Lifting off on 16 November 1973, astronauts Jerry Carr, Ed Gibson, and Bill Pogue set an 84-day endurance record for Americans in space. The many lessons learned included how crews in space interact scheduling activities with the mission control center and principal investigator scientists on the ground. The result allowed the astronauts to finish tasks, transition, and reflect on their work, resulting in a productive mission and improved processes for future operations aboard the International Space Station.

After the mission, Skylab 4 science pilot Gibson gave his view on how the Apollo era led to the first lunar landings and Skylab: "Apollo was really a great program that required us to develop new technologies," he said. "That put us in a more competitive position. What we got back from it economically was at least two to three times what we put into it."

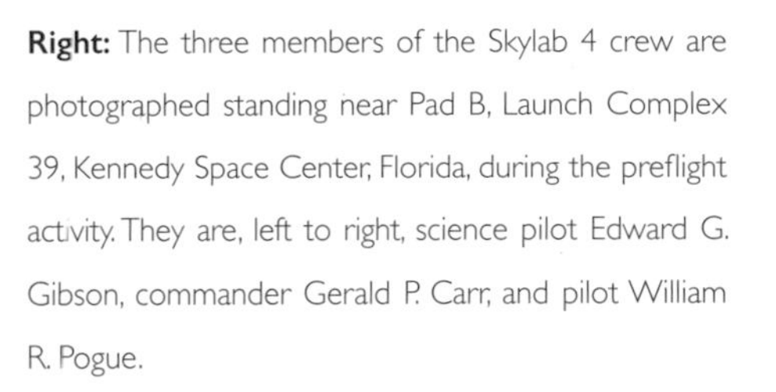

Right: The three members of the Skylab 4 crew are photographed standing near Pad B, Launch Complex 39, Kennedy Space Center, Florida, during the preflight activity. They are, left to right, science pilot Edward G. Gibson, commander Gerald P. Carr, and pilot William R. Pogue.

Opposite: Skylab 4 commander Gerald Carr jokingly demonstrates weight training in zero gravity as he balances fellow astronaut William Pogue, the mission's pilot, upside down on his finger.

[Photo and caption credits: NASA]

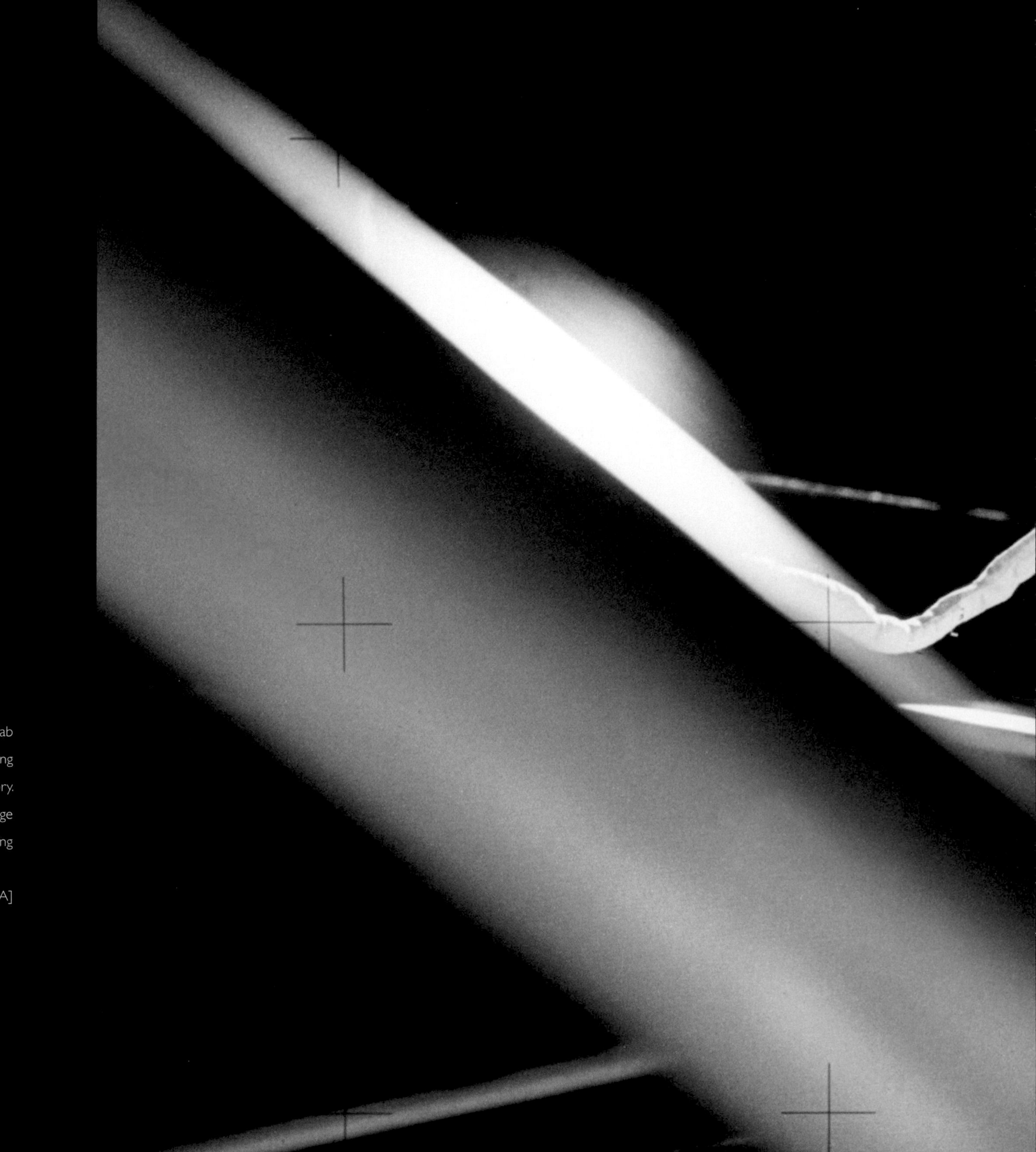

Right: In November 1973, the second crewed Skylab mission splashed down in the Pacific Ocean following a successful 59-day mission in the orbiting laboratory. Here, astronaut Owen K. Garriott retrieves an image experiment from the Apollo Telescope Mount during an extravehicular activity.

[Photo and caption credit: NASA]

Above: Astronauts on a space walk and the International Space Station with a view of Earth.
[Photo and caption credit: NASA]

International Space Station

After Skylab, NASA engineers began plans for a space station to be supported by the space shuttle. As the shuttle program began to gain momentum, President Ronald Reagan asked NASA to build a permanently crewed Earth-orbiting space station during his 1984 State of the Union address. Over time, the project evolved into the International Space Station Program supported by 15 nations around the world.

The first element of the International Space Station, the Zarya control module, was launched by the Russian space agency on 20 November 1998. *Zarya* is Russian for "dawn". The crew of Space Shuttle Endeavour soon attached the US unity node, beginning assembly of the largest spacecraft ever built. NASA and a host of other nations continue to use the space station to learn more about living and working in space. Lessons learned will help as astronauts build an outpost on the Moon and, eventually, travel to Mars.

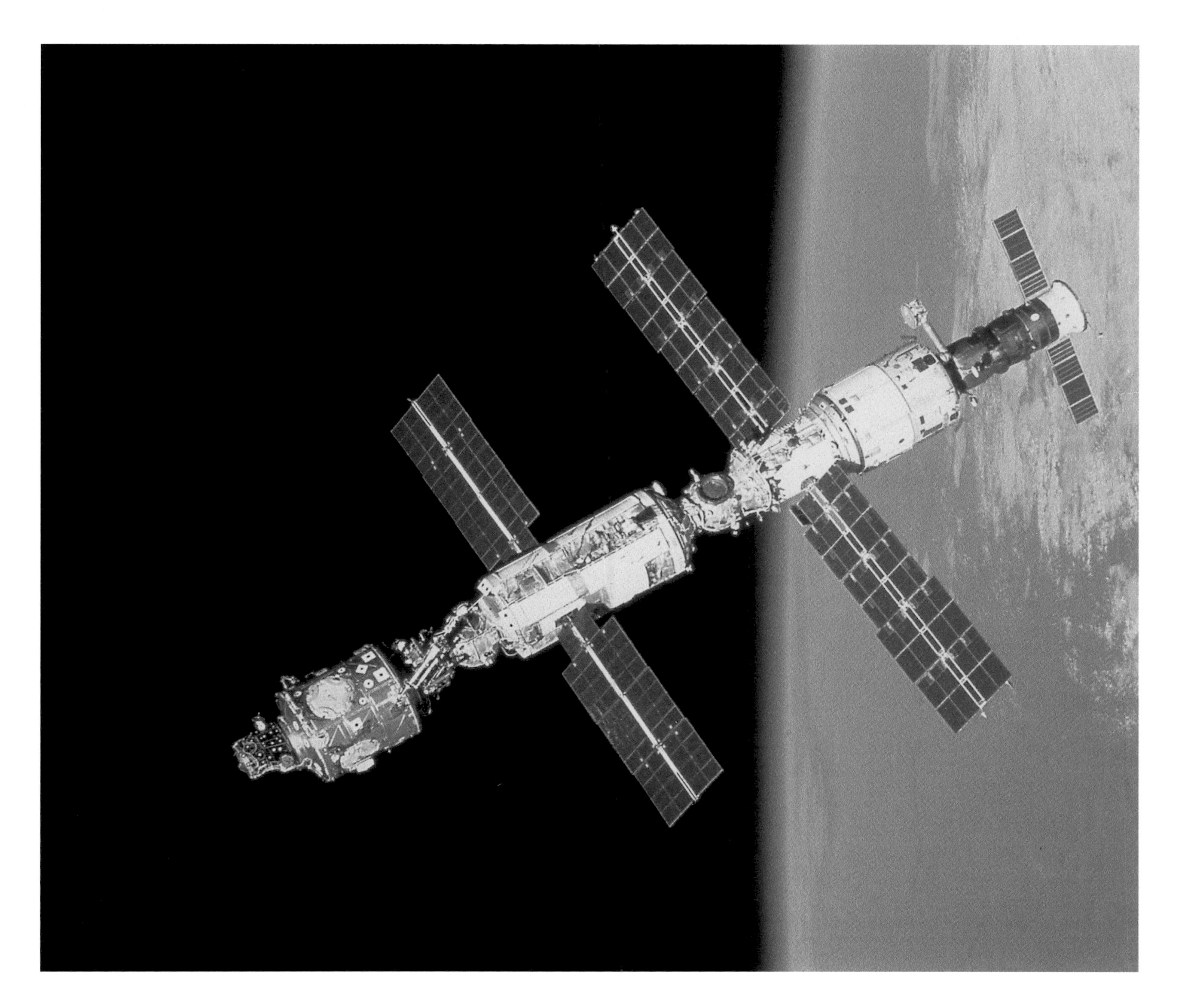

Above: This image of the International Space Station in orbit was taken from Space Shuttle Endeavour prior to docking. Most of the station's components are clearly visible in this photogaph. They are the Node 1 or Unity module docked with the functional cargo block or Zarya (top) that is linked to the Zvezda service module. The Soyuz spacecraft is at the bottom.

The Size of the Space Station

The International Space Station's volume is roughly equal to a five-bedroom house or two Boeing 747 jetliners. It supports crews of seven people plus occasional visitors. The length from the edges of the station's solar arrays stretches the length of an American football field, including the end zones. It has laboratories provided by the United States, Russia, Japan, and the European Space Agency. Airlocks opening to the outside allow astronauts to emerge for spacewalks. Mounted on the outside of the space station are robotic arms used to aid construction. The arms also move astronauts around during spacewalks, as others operate science experiments.

Right: In this artist's depiction of mission STS-88, the first space shuttle assembly flight for the International Space Station which launched in December 1998, Space Shuttle Endeavour prepares to capture the functional cargo block (FGB) using the shuttle's mechanical arm. The shuttle carried the first United States-built component for the station, a connecting module called Node 1, and attached it to the already orbiting FGB, which supplies early electrical power and propulsion. The FGB, Zarya, launched two weeks earlier on a Russian Proton rocket from the Baikonur Cosmodrome, Kazahkstan. Once the FGB was captured using the mechanical arm, astronaut Nancy J. Currie maneuvered the arm to dock the FGB to the conical mating adapter at the top of Node 1 in the Shuttle's cargo bay. In ensuing days, three extravehicular activities by astronauts Jerry L. Ross and James H. Newman were performed to make power, data, and utility connections between the two modules.

[Photo and caption credit: NASA]

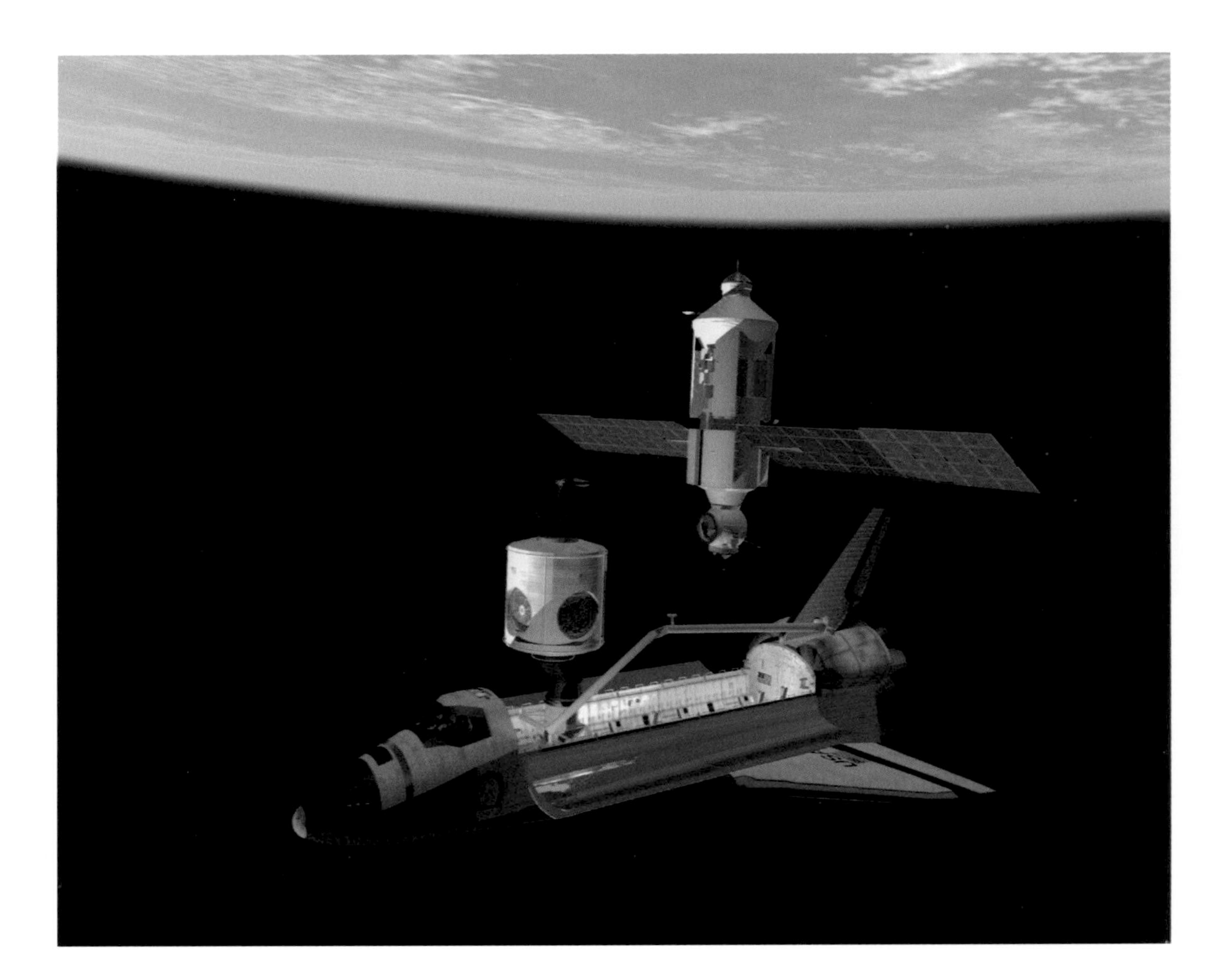

Above: The mated Russian-built Zarya and US-built Unity modules are backdropped against the blackness of space and Earth's horizon shortly after leaving Endeavour's cargo bay.

[Photo and caption credit: NASA]

The Space Station Era

As more elements were added, the station was ready for its first crew that arrived on 2 November 2000. The space station has been continuously staffed ever since, with construction completed by crews of the final space shuttle missions in 2011.

Space station crewmembers perform research that can only be done in the microgravity environment of space. Since the space station has been staffed for more than 20 years, engineers are learning how to keep a spacecraft working well over longer periods.

Many of the results of the research includes studies into what happens to the human body when astronauts live and work in microgravity over long periods of time. This will be crucial as NASA plans missions to destinations such as Mars, a round trip that will take longer than a year.

Research in life sciences, materials processing, and observation of the terrestrial climate results in what NASA refers to as working "off the Earth for the Earth," benefiting life on this planet.

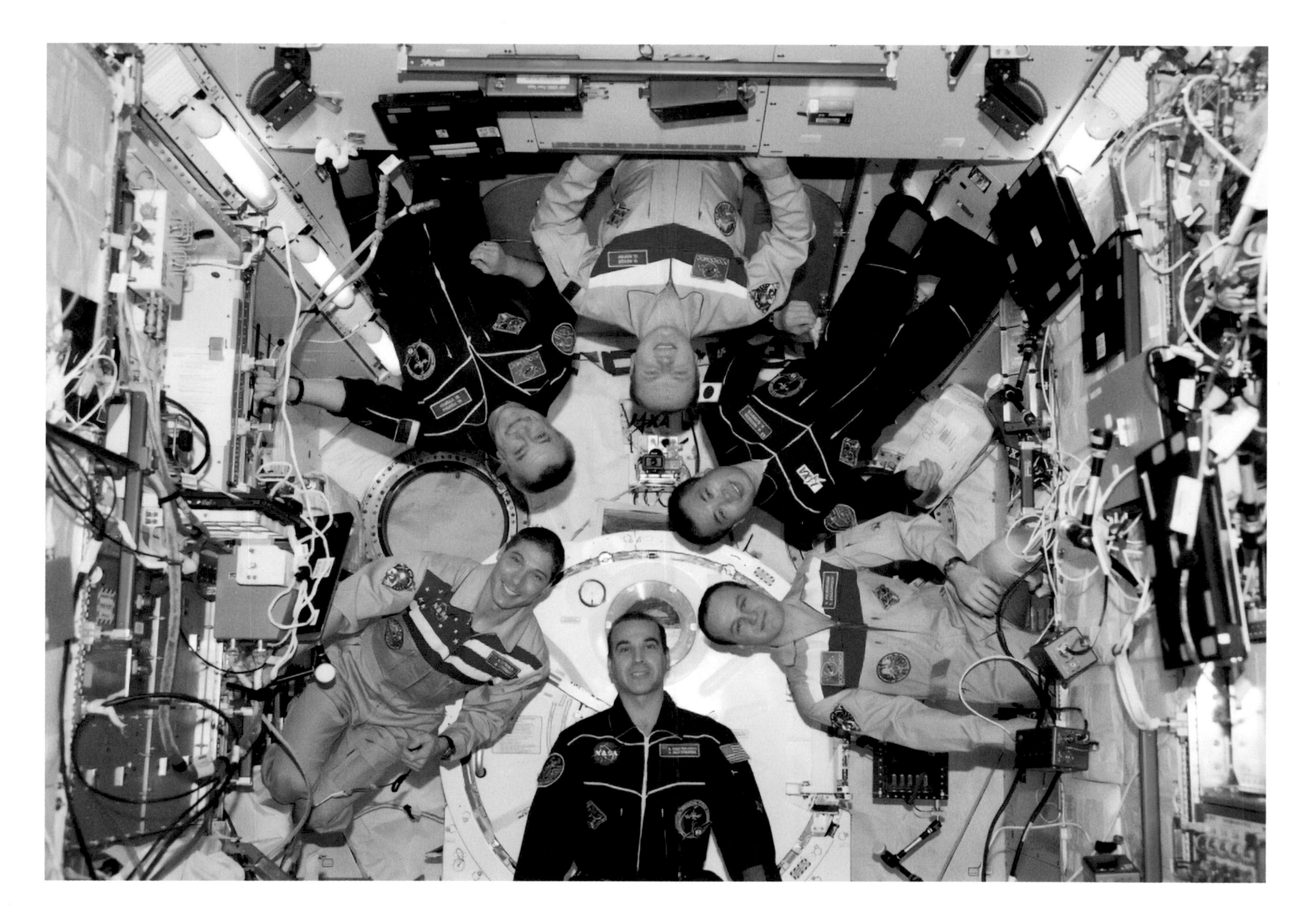

Above: Expedition 38 crew members pose for an in-flight crew portrait in the Kibo laboratory of the International Space Station on 22 February 2014. Pictured (clockwise from top center) are Russian cosmonaut commander Oleg Kotov, Japan Aerospace Exploration Agency astronaut Koichi Wakata, Russian cosmonaut Sergey Ryazanskiy, NASA astronauts Rick Mastracchio and Mike Hopkins, and Russian cosmonaut Mikhail Tyurin, all flight engineers.

[Photo and caption credit: NASA]

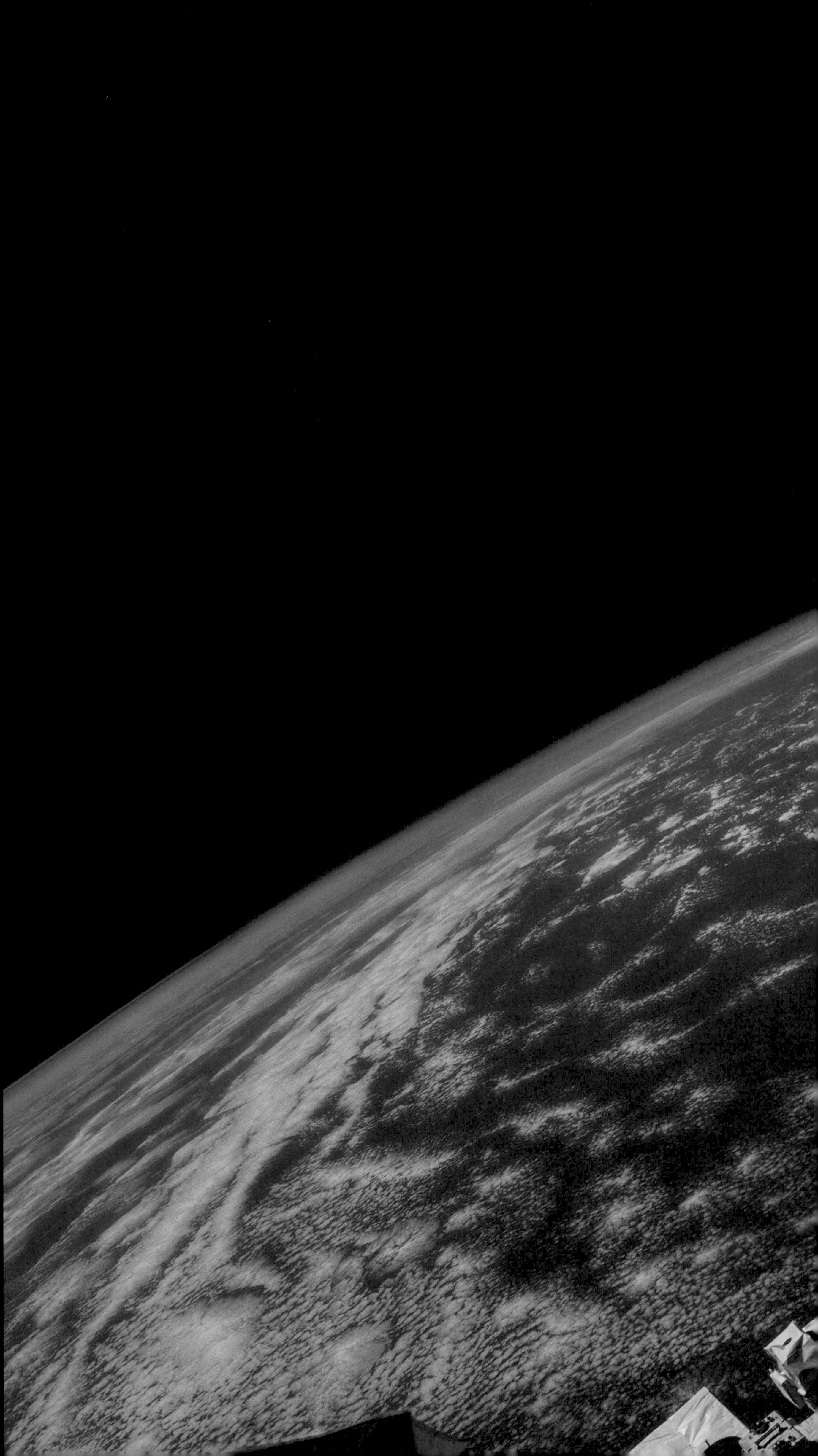

Right: Japan's Kibo laboratory module is pictured as the International Space Station orbits 267 miles above the Pacific Ocean.

[Photo and caption credit: NASA]

JAPAN

"By denying scientific principles, one may maintain any paradox."

Galileo Galilei

Left: ESA astronaut Matthias Maurer is pictured outside the International Space Station during a spacewalk.
[Photo and caption credit: NASA]

I get the feeling that, after decades of Mars sci-fi movies, photos from afar, and probes and rovers, within our lifetime a human being will indeed land on the red planet. While the Moon is only 238,900 miles away, Mars—at its closest orbit—is about 33.9 million miles away and would take around seven months to reach…if everything goes well. I'm absolutely fascinated with this planet since there are so many conflicting perspectives and still so many unknowns. From creating an earth-like atmosphere based on the detonation of atomic bombs to the potential water supply at the poles. However, the questions that boggle my mind include the psychological effects of the actual journey to and from Mars, which territory(s) will impose their rules, and how politics, religion, and commercialism might influence its future. As Mars is certainly the next great human destination, this next chapter helps us to better understand this fascinating red planet.

esa

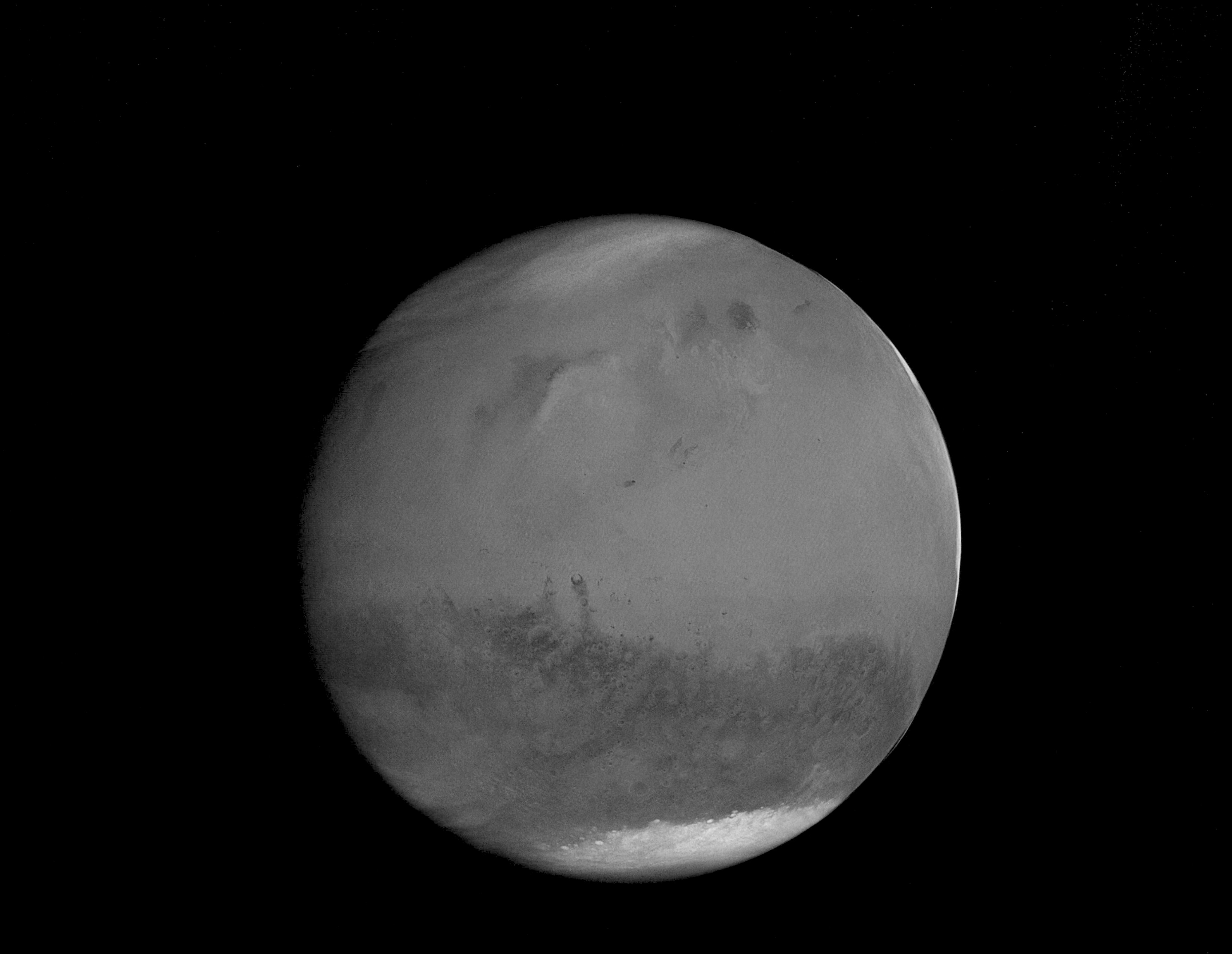

Above: This photograph of Mars is the first one taken that shows its true colour and has been achieved by using orange (red), green, and blue filters. It was taken by the OSIRIS instrument on the European Space Agency's Rosetta spacecraft during its February 2007 flyby of the planet.

Why Mars?

The planet Mars has always been an enigma. In 1939, Americans were already jittery over events abroad leading to a global conflict. They were further frightened by Orson Wells' 1938 radio broadcast of H.G. Wells' classic book, *War of the Worlds* (1897) in which Martians fictitiously invaded Earth. However, beginning in the 1960s, an armada of robotic spacecraft have sought to unravel the mysteries by flying by, orbiting, landing on and exploring the Red Planet, searching for signs of simple, molecular life. NASA is now making plans for the first human expedition there to answer long-standing questions about the far-away world.

At an average distance of 442-million-miles from Earth, Mars is a terrestrial planet with an atmosphere, climate, and geology similar to Earth. Scientists and astronomers believe exploration of the Red Planet may answer many questions about the solar system.

NASA's exploration of Mars began on 28 November 1964 when Mariner 4 was launched as the first successful mission to the Red Planet. The spacecraft returned the first photographs of the surface from a range of 6,118 miles, showing that it was pockmarked with craters similar to those on the Moon.

The first spacecraft to enter orbit around another planet was Mariner 9. The robotic spacecraft reached Mars on 14 November 1971, sending back 7,329 images during its 16-month mission.

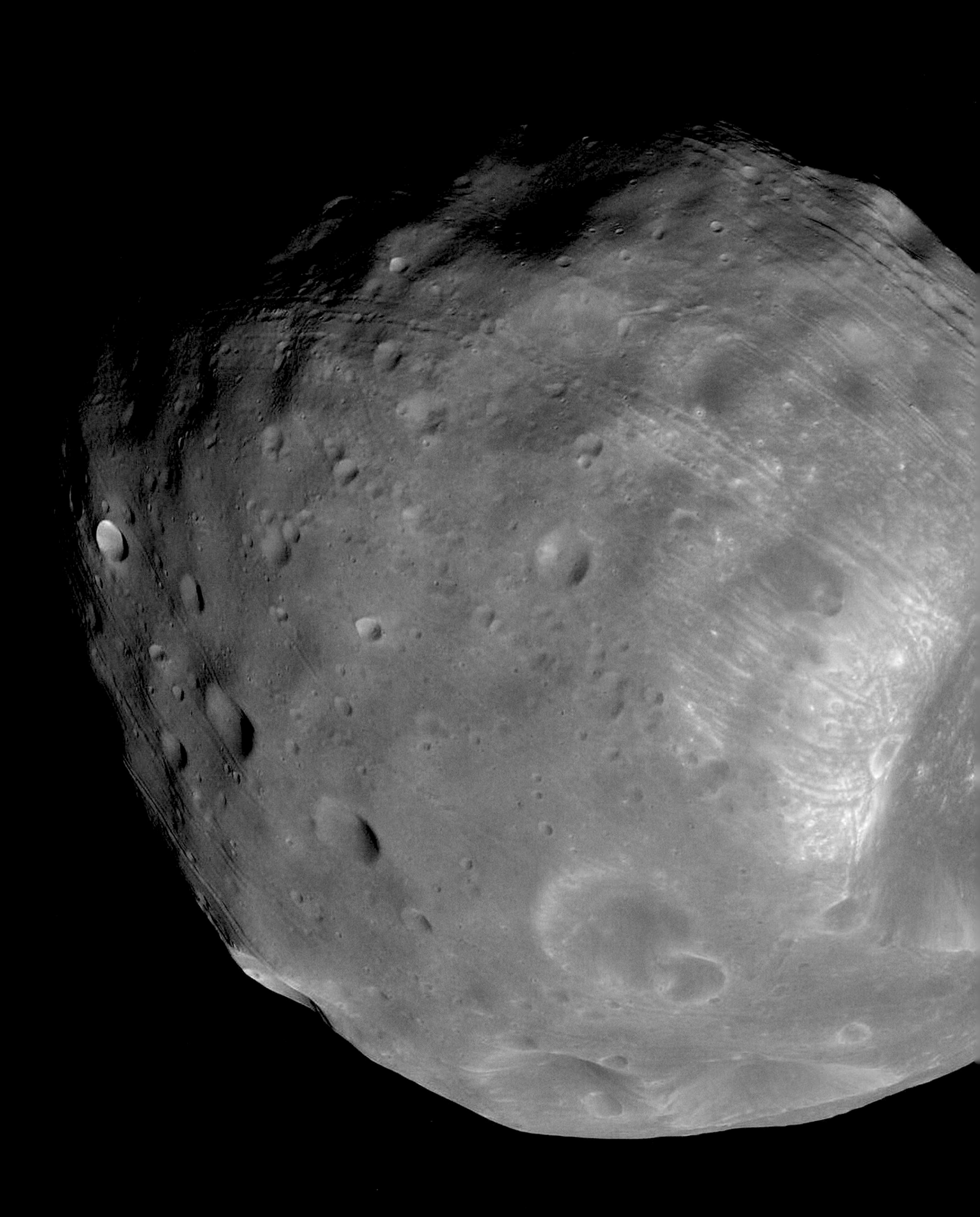

Phobos and Deimos

Right: Mars, named for the Roman god of war, has two tiny moons—Phobos and Deimos—whose names are derived from the Greek for fear and panic. These Martian moons may well be captured asteroids originating in the main asteroid belt between Mars and Jupiter or perhaps from even more distant reaches of the solar system. The larger moon, Phobos, is a cratered, asteroid-like object that orbits so close to Mars that gravitational tidal forces are dragging it down. In 100 million years or so, Phobos will likely be shattered by stress caused by the relentless tidal forces, the debris forming a decaying ring around Mars. Deimos, the smaller of the two moons, has a smooth surface due to a blanket of fragmental rock or regolith, except for the most recent impact craters.

[Photo and caption credit: NASA]

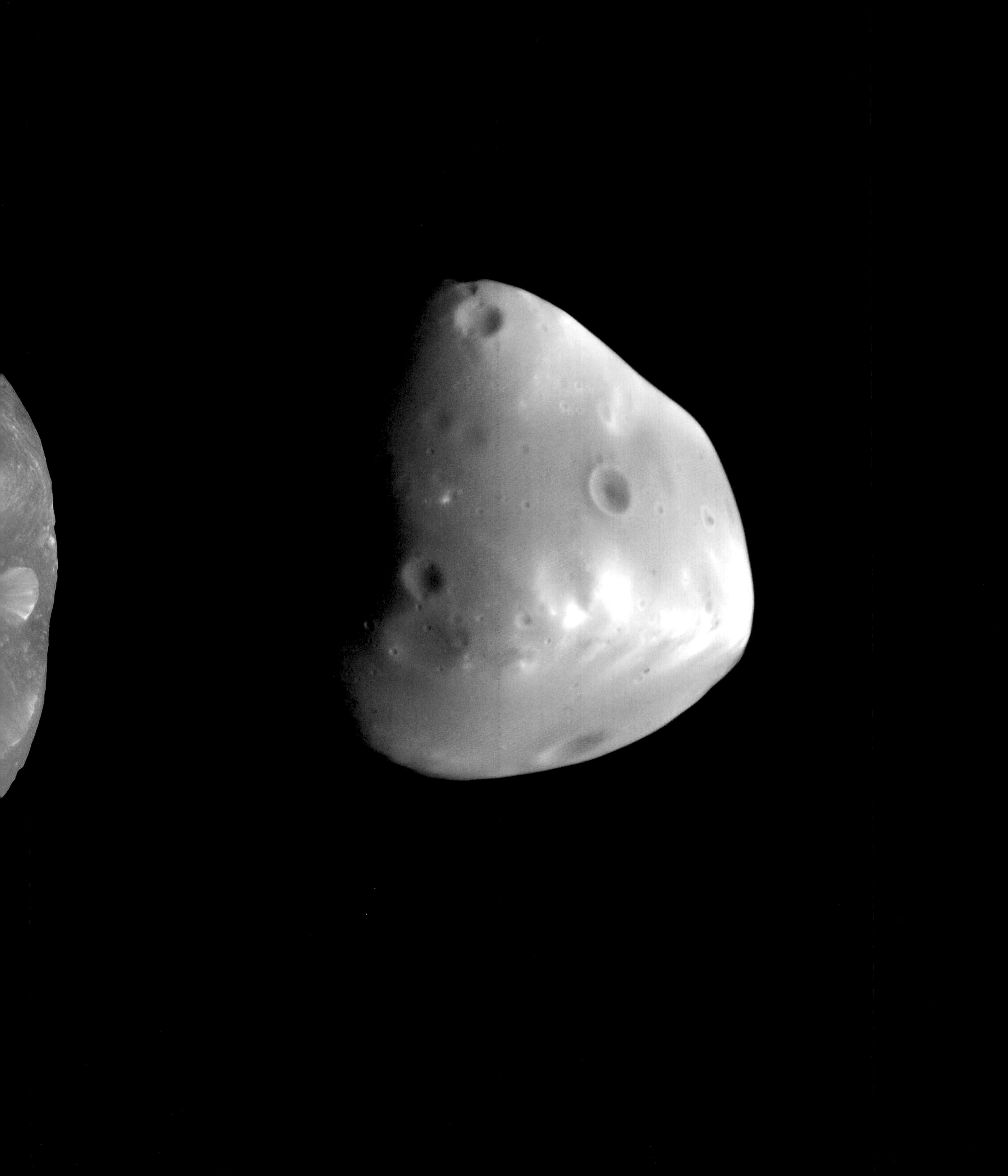

NASA's Viking Project

In August and September 1975, the United States began an ambitious Mars mission with twin Viking spacecraft launched three weeks apart. The Vikings included spacecraft designed to both orbit and land on Mars.

After Vikings 1 and 2 entered orbit around Mars, the landers each separated from their respective spacecraft and landed on the planet's surface. The first Viking touched down on 20 July 1976 on a slope of a plain called Chryse Planitia, making it the first successful Mars lander. Viking 2 landed on 3 September 1976 in a large plain called Utopia Planitia.

While the orbiters took thousands of images of the Martian surface, the landers collected science data and conducted geology investigations. The data suggests that the environment of Mars is what is termed, "self-sterilizing." The solar ultraviolet radiation, coupled with the dry soil and the oxidizing nature of the soil chemistry, thwarts formation of organisms.

Above: Photograph of a full-scale Viking lander model.

[Photo and caption credit: NASA/JPL]

Right: The Viking 2 lander and orbiter was launched from a Titan IIIE-Centaur. They separated when they reached the planet and the lander made its treacherous descent to the Martian surface. The orbiter ceased sending telemetry back to Earth shortly after separation. Engineers scrambled to turn on the tape recorder so that any images would be saved and sent when the antenna was working properly. Nine hours after the originally scheduled viewing, images came through at 16 kilobits per second. Operations of Viking 2 ceased in 1980.

[Photo and caption credits: NASA]

First Mars Rovers

Two decades after the first successful landers, NASA launched its first mission to place a robotic rover on the Martian surface. On 4 July 1997, Mars Pathfinder touched down with a base station. Additionally, a small 23-pound, six-wheeled rover named "Sojourner" became the first remote-controlled vehicle to move around on another planet. The rover analyzed nearby rocks, determining those in the surrounding area were formed by the evaporation of floodwaters.

As Sojourner examined nearby objects, engineers commanded it to use an Alpha Proton X-Ray Spectrometer to analyze a stone that resembled the face of a bear. Scientists nicknamed the rock, "Yogi," after the popular children's cartoon character, Yogi Bear. Data returned from the sample indicated that the rock contained little quartz and was similar to basalts on Earth.

Right: Mars Pathfinder Sojourner.

Spirit and Opportunity

The Mars Exploration Rover program involved NASA launching Mars rovers named Spirit and Opportunity in June and July 2003. Designed to operate for a little more than three months, Spirit was operational more than six years anc Opportunity exceeded its expected life span by 14 years. Between the two, they traversed 33 miles across the Martian surface. Discoveries made by the twin rovers included the possibility that in the distant past, at least one area of Mars was wet enough for a long enough time to have supported microbial life.

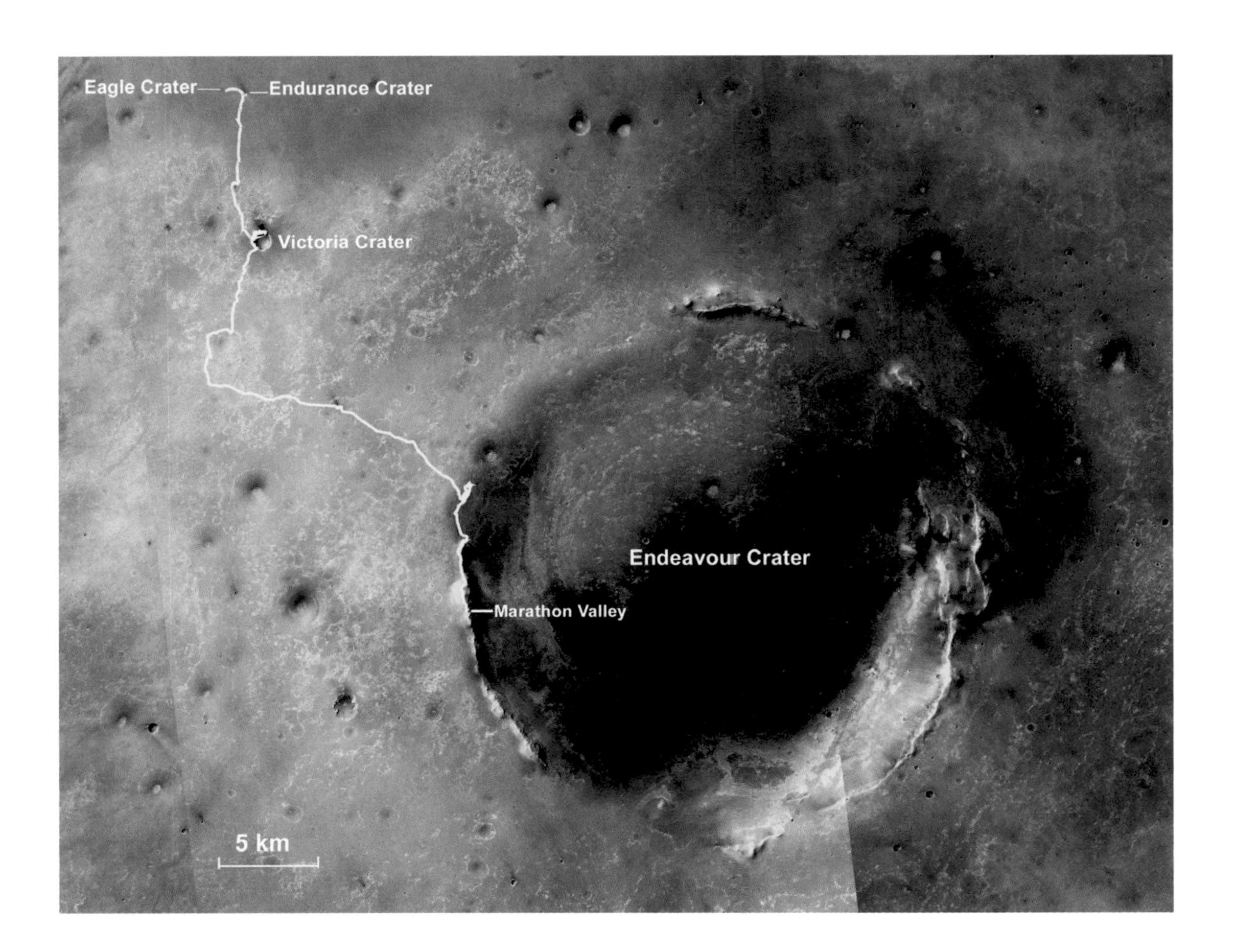

Left: The gold line on this image shows Opportunity's route from the landing site inside Eagle Crater, in upper left, to its location after the Sol 3968 drive. The mission has been investigating on the western rim of Endeavour Crater since August 2011. This crater spans about 14 miles (22 kilometers) in diameter. The mapped area is all within the Meridiani Planum region of equatorial Mars, which was chosen as Opportunity's landing area because of earlier detection of the mineral hematite from orbit. North is up.

[Photo and caption credits: NASA/JPL-Caltech]

"We are all…children of this universe. Not just Earth, or Mars, or this system, but the whole grand fireworks. And if we are interested in Mars at all, it is only because we wonder over our past and worry terribly about our possible future."

Ray Bradbury, author and screenwriter

Above: Martian landscape as seen by Mars Rover.

[Photo credit: Alamy Stock Photo]

Phoenix Mars Lander

Launched on 4 August 2007, NASA's Phoenix lander was an effort to find resources on Mars that can be used by future astronaut explorers who will need to live off the land. The spacecraft touched down on 25 May 2008 in an area near the ice-rich soil on the Red Planet's arctic surface. Phoenix used a robotic arm to collect soil samples for analysis by the spacecraft's systems, confirming that there is water ice on Mars.

The presence of water is a significant find. The elements in water—hydrogen and oxygen—would help sustain a human presence on Mars and the basics for rocket fuel for a return trip to Earth. In addition, the Phoenix lander checked soil samples for evidence of simple forms of life in either the past or present.

Phoenix was equipped with a six-and-a-half foot mast that extended with a stereo camera. The instrument returned high-resolution images providing insights into the area's geology. The multi-spectral capability of the cameras aided scientists in confirming the types of minerals present near the landing site. The lander also scanned the Martian atmosphere up to an altitude of 12.4 miles. The findings provided information on dust plumes prevalent on the Red Planet along with the duration and movement of clouds and fog. In October 2008, the Phoenix lander's solar panels were unable to receive enough sunlight to keep batteries charged, but the spacecraft had already exceeded expectations.

Above: The Phoenix Mars lander launched on 4 August 2007 on Delta II from Cape Canaveral Air Force Station in Florida. It landed on 25 May 2008 at Vastitas Borealis, the arctic plains of Mars. It's mission ended on 2 November 2008.

[Photo and caption credit: NASA]

Curiosity Rover

The Mars Science Laboratory with its Curiosity rover was NASA's first large-scale robotic mission to study Mars. The size of a compact car, Curiosity is double the proportions of Spirit and Opportunity. Launched on 26 November 2011, the trip took nine months to complete, landing in Gale Crater on 6 August 2012.

As was the case with earlier landers, Curiosity's instruments confirmed that the Martian climate included habitable environments. While controlled by engineers at NASA's Jet Propulsion Laboratory in California, the rover moved around Gale Crater using its seven-foot arm to collect rocks, soil, and air samples for analysis by onboard instruments.

Below: This low-angle self-portrait of NASA's Curiosity Mars rover shows the vehicle at the site from which it reached down to drill into a rock target called "Buckskin" on lower Mount Sharp.

[Photo and caption credit: NASA]

Above: Gale Crater was formed when a meteor hit Mars in its early history, about 3.5 to 3.8 billion years ago, with the impact punching a hole in the terrain. The explosion ejected rocks and soil that landed around the crater. Scientists chose Gale Crater as the landing site for Curiosity because it has many signs that water was present over its history.

[Photo and caption credit: NASA]

Curiosity's Clever Toolkit

The rover's toolkit included 17 cameras, a laser used to vaporize and study regolith, as well as a drill to check and grind rock samples. Curiosity was designed to analyze the powdered samples, measuring the chemical "fingerprints" in various minerals and providing insights into their composition and history.

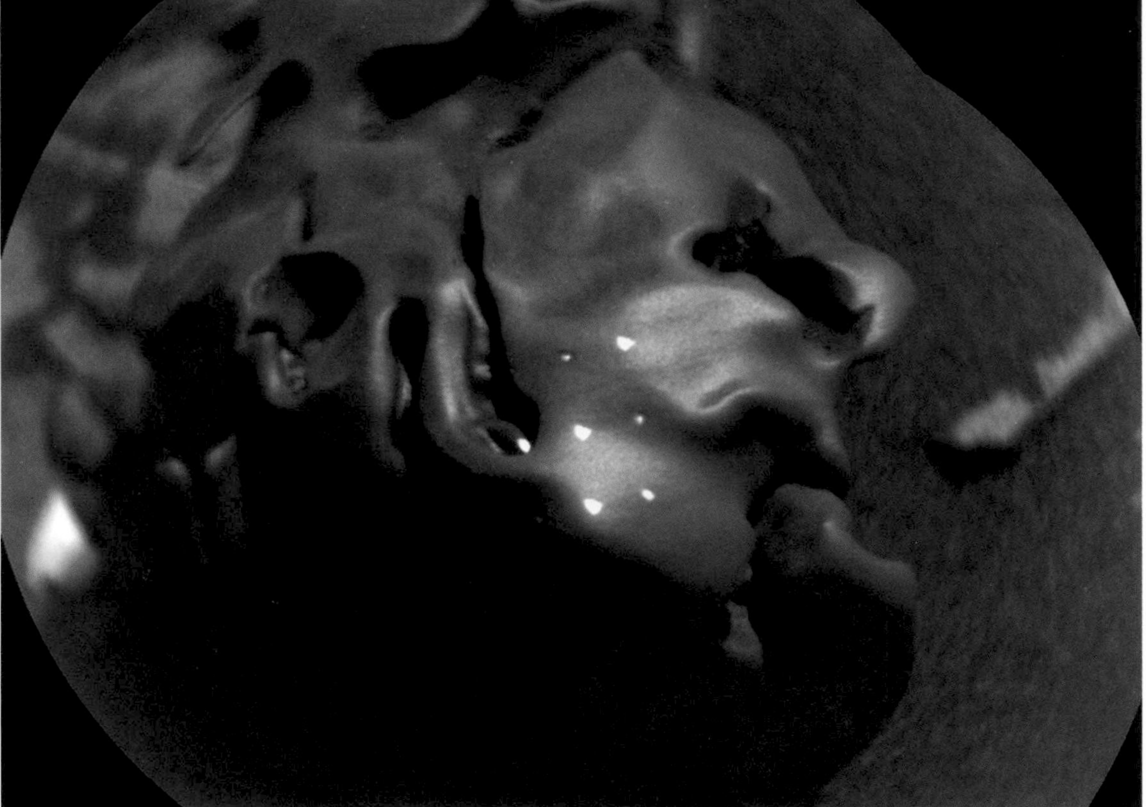

Top: This view of a Martian rock slab called "Old Soaker," which has a network of cracks that may have originated in drying mud, comes from the mast camera (Mastcam) on NASA's Curiosity Mars rover. The location is within an exposure of Murray formation mudstone on lower Mount Sharp inside Gale Crater. Mud cracks would be evidence of a time more than 3 billion years ago when dry intervals interrupted wetter periods that supported lakes in the area. Curiosity has found evidence of ancient lakes in older, lower-lying rock layers and also in younger mudstone that is above Old Soaker.
Bottom: This image of a dark, golf-ball-size object was taken by the chemistry-and-camera instrument (ChemCam) on NASA's Curiosity Mars rover. By firing laser pulses to determine the chemical elements in the object's composition, the rover's analysis confirmed that this object, informally named "Egg Rock," is an iron-nickel meteorite.

[Photo and caption credits: NASA]

Above: This image was taken by the Left Navigation Camera onboard NASA's Mars rover Curiosity on Sol 3143.

[Photo and caption credit: NASA]

As was the case with most Mars landers, one objective was to answer the long-standing question, Does life exist on the Red Planet? Many of Curiosity's instruments have been utilized to seek indications of whether any portion of Mars is habitable for any form of microbial life. Early in 2022, researchers determined that some of the samples studied by the rover are rich in types of carbon that can help support biological processes on Earth. More study will be needed, however, to confirm whether or not this finding points to microbes such as ancient bacteria existing on Mars.

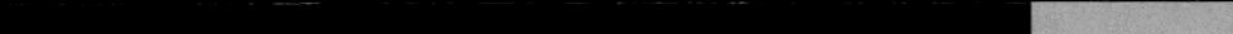

Mount Sharp, Mars

Right: A view from the "Kimberley" formation on Mars taken by NASA's Curiosity rover. The strata in the foreground dip towards the base of Mount Sharp, indicating flow of water toward a basin that existed before the larger bulk of the mountain formed. The colors are adjusted so that the rocks look approximately as they would if they were on Earth, to help geologists interpret the rocks. This "white balancing" to adjust for the lighting on Mars overly compensates for the absence of blue on Mars, making the sky appear light blue and sometimes giving dark, black rocks a blue cast. This image was taken by the mast camera (Mastcam) on Curiosity on the 580th Martian day, or sol, of the mission.

[Photo and caption credit: NASA/JPL-Caltech/MSSS]

Studying the Inner Space of Mars

While robotic spacecraft continue to study Mars from orbit and the surface, NASA's InSight (which stands for Interior Exploration using Seismic Investigations Geodesy and Heat Transport) is analyzing the Red Planet from deep inside its interior. InSight launched on 5 May 2018. As is the case with all NASA planetary probes, InSight is managed by the agency's Jet Propulsion Laboratory with many of its scientific instruments designed and fabricated by the European Space Agency (ESA).

After a six-month trip, the lander arrived on 26 November 2018, landing at Elysium Planitia. InSight drilled into the interior crust, mantle, and core that scientists believe formed 4.5 billion years ago. It is also designed to measure tectonic activity and meteorite impacts. Studying Mars' interior is expected to aid in determining how future astronauts will be able to live off the land, as well as answer questions regarding the formation of the solar system.

Right: This illustration shows NASA's InSight spacecraft with its instruments deployed on the Martian surface. **Opposite:** Artist's rendition showing the inner structure of Mars. The topmost layer is known as the crust, underneath it is the mantle, which rests on a solid inner core.

[Picture and caption credits: NASA/JPL-Caltech]

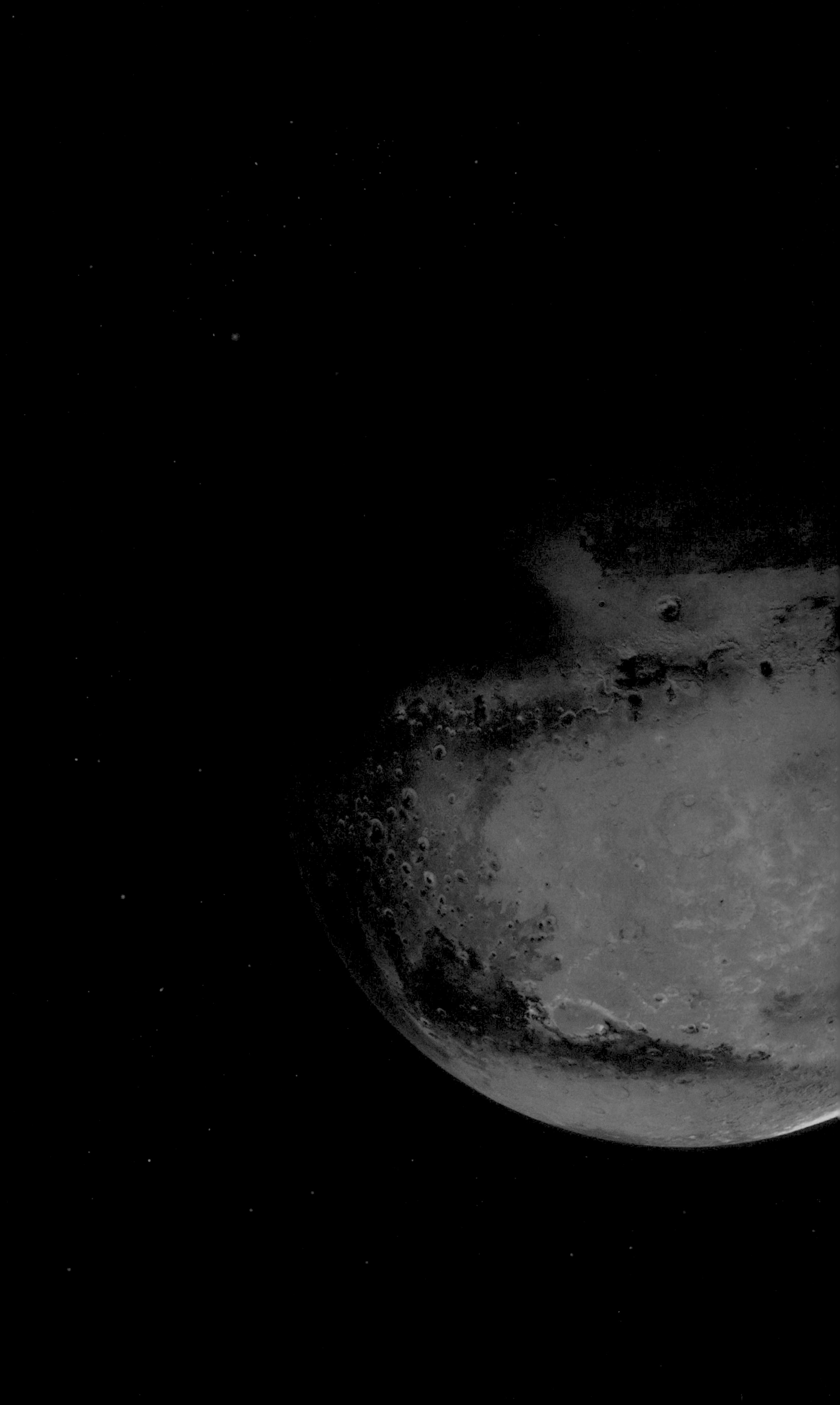

Right: This illustration shows NASA's Mars 2020 spacecraft carrying the Perseverance rover as it approaches Mars. Hundreds of critical events must execute perfectly and exactly on time for the rover to land on Mars safely. Solar panels powering the spacecraft are visible on the cruise state at the top. The cruise stage is attached to the aeroshell, which encloses the rover and descent stage. Entry, descent and landing, or "EDL," begins when the aeroshell reaches the top of the Martian atmosphere, traveling nearly 12,500 mph (20,000 kph). It ends about seven minutes later, with Perseverance stationary on the Martian surface.

[Picture and caption credit: NASA/JPL-Caltech]

Perseverance Rover

NASA's Mars 2020 Perseverance rover launched on 30 July 2020, continuing an exploration of Mars that has been ongoing for more than half a century. Perseverance incorporates designs to reduce costs and risks based on the approach taken for the earlier Curiosity rover.

The spacecraft landed on Mars on 18 February 2021 and immediately began beaming back detailed photographs from Jezero Crater. The site was chosen because scientists believe the area was once flooded and was the location of an ancient river delta.

Perseverance is also checking out technology for collecting oxygen from the atmosphere of Mars that contains 96 percent carbon dioxide. This could aid in planning for the use of

Below: This map of Mars shows the landing site for NASA's Perseverance rover in relation to those of previous successful Mars missions.
[Photo and caption credit: NASA]

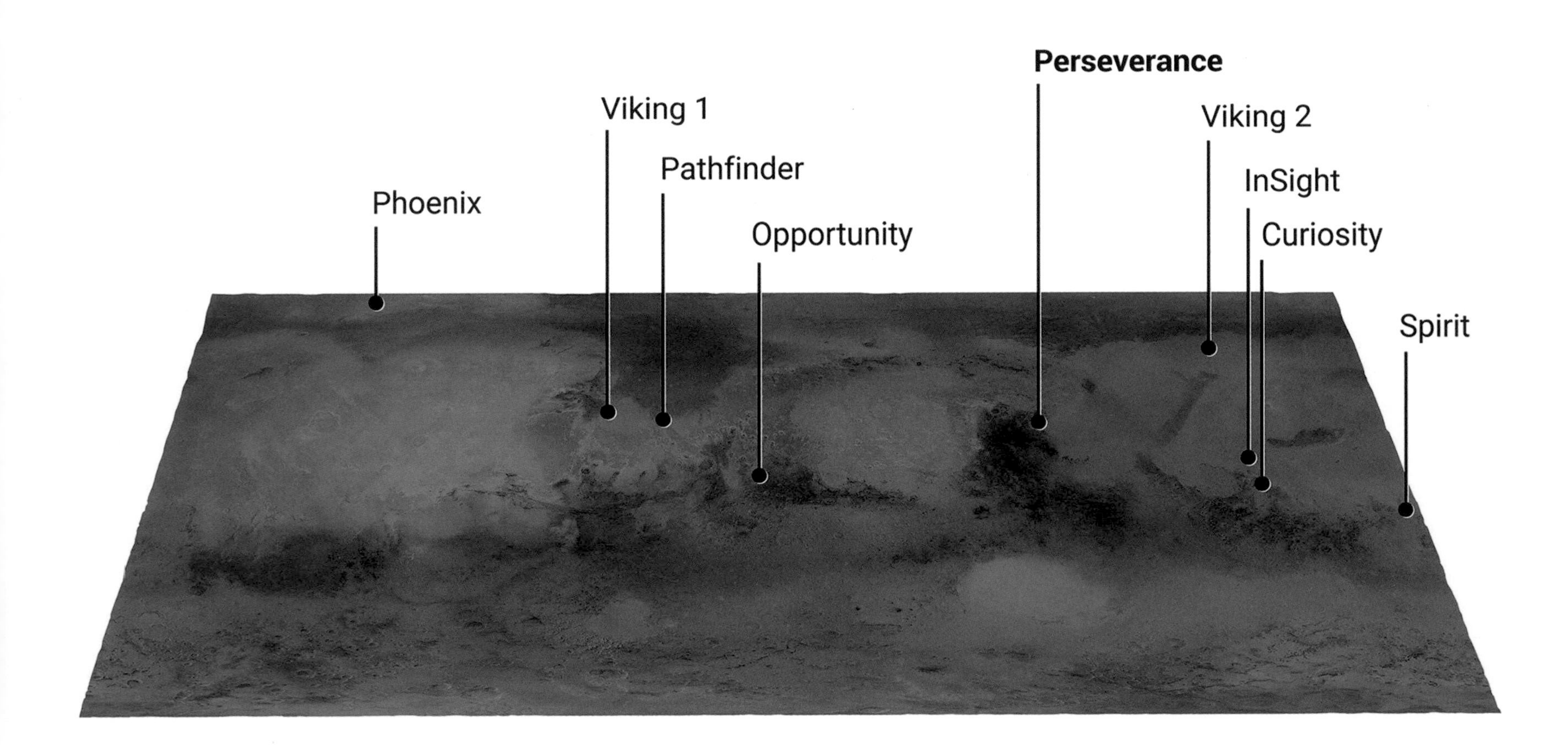

Martian natural resources to support astronauts, as well as developing ways to design life support systems, transportation, and other processes for living and working on the Red Planet. The spacecraft has a mobility system allowing it to travel farther than previous rovers. Perseverance is also designed to gather and store samples from the Martian surface, possibly leading to a future mission to retrieve the soil for study on Earth.

Above: This high-resolution still image is part of a video taken by several cameras as NASA's Perseverance rover touched down on Mars on 18 February 2021. A camera aboard the descent stage captured this shot. [Photo and caption credit: NASA]

NASA and ESA are working to develop a spacecraft that could retrieve samples collected by Perseverance. The NASA-led sample retrieval lander would travel to Mars, picking up the collected soil and launch the package into orbit around Mars. An ESA-led Earth return orbiter would rendezvous with the sample-carrying spacecraft and return it to Earth.

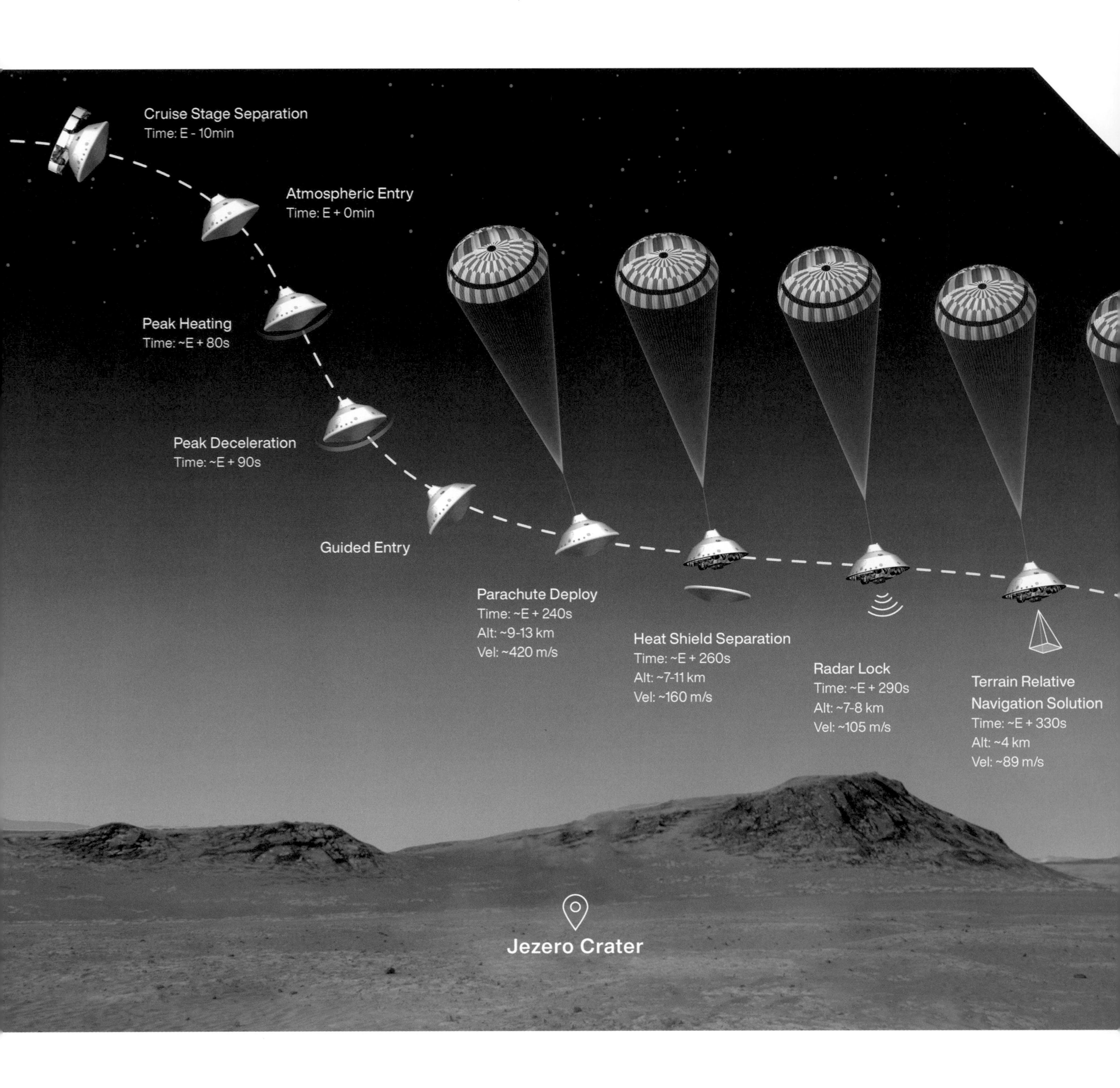
Cruise Stage Separation
Time: E - 10min
Atmospheric Entry
Time: E + 0min
Peak Heating
Time: ~E + 80s
Peak Deceleration
Time: ~E + 90s
Guided Entry
Parachute Deploy
Time: ~E + 240s
Alt: ~9-13 km
Vel: ~420 m/s
Heat Shield Separation
Time: ~E + 260s
Alt: ~7-11 km
Vel: ~160 m/s
Radar Lock
Time: ~E + 290s
Alt: ~7-8 km
Vel: ~105 m/s
Terrain Relative
Navigation Solution
Time: ~E + 330s
Alt: ~4 km
Vel: ~89 m/s
Jezero Crater

Coming in for a Landing

Perseverance had incredibly challenging terrain to tackle when it came to landing on Mars. The designated landing place, Jezero Crater, is 28-miles wide and made up of steep cliffs, boulder fields, sand dunes and small impact craters. Thanks to EDL technologies, such as Range Trigger, which works out when Perseverance's parachute can be deployed based on its navigation position, the rover was able to land safely.

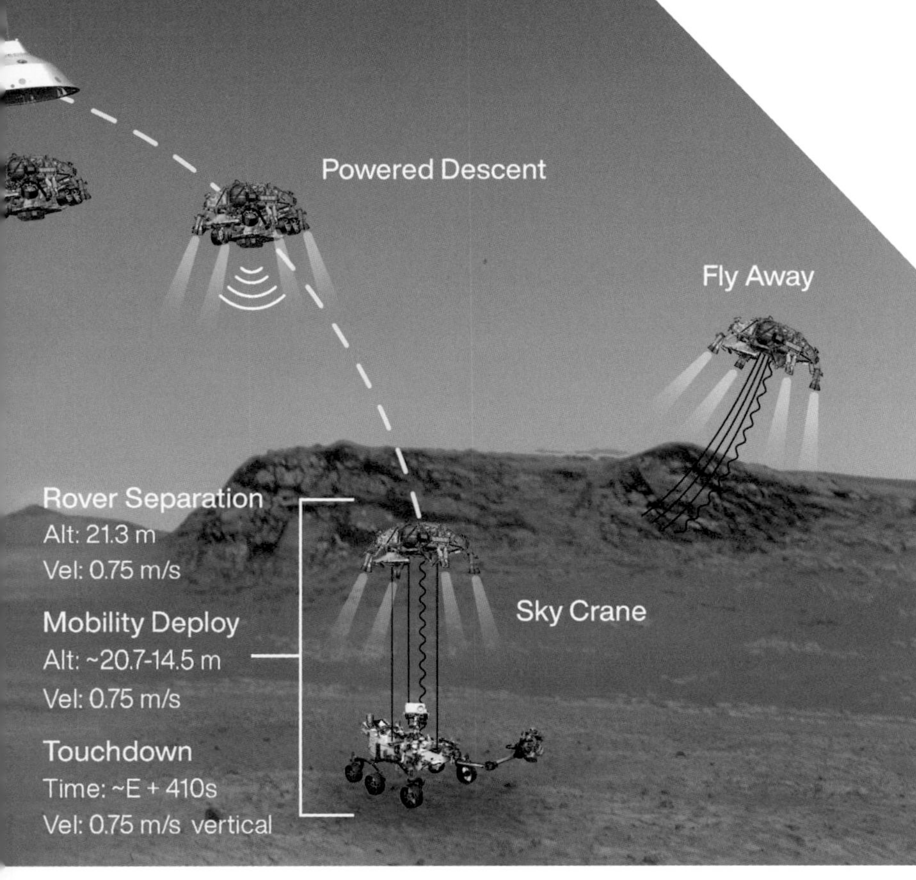

Left: This artist's concept shows the different stages that Perseverance rover goes through, from first entering the atmosphere, through guided entry, parachute deployment, the navigation procedure and finally separation and descent.

[Photo credit: NASA/JPL-Caltech]

Planetary Protection Technologies

Robotic explorers have not yet confirmed if life in any form has existed on Mars. As NASA sends more spacecraft to the surface of the Red Planet, the agency is using technology to ensure microorganisms from Earth are not introduced into the environment of another body in the solar system. As a result, the agency is using higher standards than those used for transporting hazardous substances on Earth, including sterilization processes.

Below: This image shows major components of NASA's Mars 2020 mission in the High Bay 1 clean room in Jet Propulsion Lab's (JPL) Spacecraft Assembly Facility.
[Photo and caption credit: NASA/JPL-Caltech]

Mars Ascent Vehicle

A sample retrieval lander is a first step in developing a Mars ascent vehicle, essential to return materials or astronauts. NASA is studying propulsion technology to both complete a trip to Mars and launch back for the long journey home to Earth.

Each of these robotic explorers is a precursor to human trips to Mars. As NASA gathers more information about the Red Planet and further develops the technologies to travel and live there, the effort is helping the agency to prepare for the epic journey. Flights to land on Mars, explore for months, and then return to Earth, will take more than a year. By combining lessons from the International Space Station with what robotic flights have discovered, NASA is close to making that next dream a reality.

Left: This illustration shows a concept of how the NASA Mars ascent vehicle, carrying tubes containing rock and soil samples, could be launched from the surface of Mars in one step of the Mars sample return mission.

[Picture and caption credit: NASA/JPL-Caltech]

Jezero Crater

NASA's Mars 2020 Perseverance rover was targeted to land in the Martian Jezero Crater. The process of choosing a landing site involved a collaboration of mission team members working with scientists from around the world. They carefully examined more than 60 possible locations on the planet's surface. The Jezero site was chosen because scientists believe the area was once flooded and that it was the location of an ancient river delta. The spacecraft landed on Mars on 18 February 2021 and immediately began beaming back detailed photographs from the crater.

Above: This image depicts a possible area through which the Perseverance Mars rover could traverse across Jezero Crater as it investigates several ancient environments that may have once been habitable. The route begins at the cliffs defining the base of a delta produced by a river as it flowed into a lake that once filled the crater. The path then traverses up and across the delta toward possible ancient shoreline deposits, and then climbs the 2,000-foot-high (610-meter-high) crater rim to explore the surrounding plains. About half of this traverse could be completed in Perseverance's prime mission (one Mars year, or two Earth years). For reference, the prominent crater near the center of the image is about 0.6 miles (1 kilometer) across.

[Photo and caption credit: NASA/JPL-Caltech/USGS]

"Mars tugs at the human imagination like no other planet. With a force mightier than gravity, it attracts the eye to the shimmering red presence in the clear night sky."

John Noble Wilford, science journalist

Above: This illustration shows Jezero Crater—the landing site of the Mars 2020 Perseverance rover—as it may have looked billions of years ago on Mars, when it was a lake. An inlet and outlet are also visible on either side of the lake.

[Picture and caption credit: NASA/JPL-Caltech]

Technologies for Surface Operations

Perseverance was designed to last for 90 days on Mars, although its mobility system means that it can travel farther than previous rovers, exploring more than 3 to 12 miles (5 to 20 kilometers). Surface operations are focused on two areas: navigation, which will enable Perseverance to reach areas that NASA scientists are keen to research further; and scientific investigation, whereby Perseverance will use its science instruments to find out more about the environment on Mars. The rover has been designed to gather and store samples from the Martian surface known as "depot cashing." This may lead to a future mission to retrieve the soil for study on Earth. Scientists and engineers are developing a spacecraft that could retrieve samples collected by Perseverance. The NASA-led Sample Retrieval Lander would travel to Mars, picking up the collected soil and launch the package into orbit around Mars. An Earth Return Orbiter would rendezvous with the sample-carrying spacecraft and return it to Earth.

Above: Engineers test drive the Earth-bound twin of NASA's Perseverance Mars rover for the first time in a warehouse-like assembly room at the agency's Jet Propulsion Laboratory in Southern California on 1 September 2020. This full-scale engineering version of Perseverance helps the mission team gauge how hardware and software will perform before they transmit commands to the real rover on Mars. This vehicle system test bed (VSTB) rover is also known as OPTIMISM (Operational Perseverance Twin for Integration of Mechanisms and Instruments Sent to Mars).

[Photo and caption credit: NASA]

Valles Marineris

A massive canyon called Valles Marineris extends across the Martian surface, stretching 1,900 miles (3,000 kilometers) long and 375 miles (600 kilometers) wide. As such, it is the largest known canyon in the solar system and is thought to stretch the equivalent of the width of the United States. By way of example, in contrast the Grand Canyon in the USA is 497 miles (800 kilometers) long and up to 18 miles (30 kilometers) wide. While the Grand Canyon was carved by the Colorado River, the cause of Valles Marineris remains a mystery. In recent years, geologic processes have been observed, leading some planetary scientists to theorize that the canyon may be a crack that formed as Mars cooled billions of years ago.

Ophir Chasma

Ophir Chasma was named after Ophir, a land mentioned in the Bible that was known for its riches, and is part of the Valles Marineris canyon system. Scientists believe the rocks on the ground may be volcanic magma that shoved aside objects on the canyon floor and now appear "frozen."

Top: This image of Valles Marineris was assembled from more than 100 photographs taken during the 1970s by the orbiters of Vikings 1 and 2.

Bottom: The northern part of the Valles Marineris canyon is Ophir Chasma seen in this photograph of the chasm's floor and wall. Many layers of grainy deposits are visible on the walls along with the ground marked by wind-blown ridges along with mounds of sand.

[Photo credits: NASA]

Martian Terrain

Mars is made up from different types of rocks, such as sandstone, mudstone, and basalt, and from minerals containing silicon and oxygen that were formed billions of years ago. There are a multitude of surface features, some of which can be seen in the images here.

Ejecta Blanket

This image from NASA's Mars Reconnaissance Orbiter shows the exposed bedrock of an ejecta blanket of an unnamed crater in the Mare Serpentis region of Mars. Ejecta, when exposed, are truly an eye-opening feature, as they reveal the sometimes exotic subsurface, and materials created by impacts (close-up views). This ejecta shares similarities to others found elsewhere on Mars, which are of particular scientific interest for the extent of exposure and diverse colors. The colors observed in this picture represent different rocks and minerals, now exposed on the surface. Blue in HiRISE (the camera on board the Mars Reconnaissance Orbiter) infrared color images generally depicts iron-rich minerals, like olivine and pyroxene. Lighter colors, such as yellow, indicate the presence of altered rocks.

[Photo and caption credit: NASA]

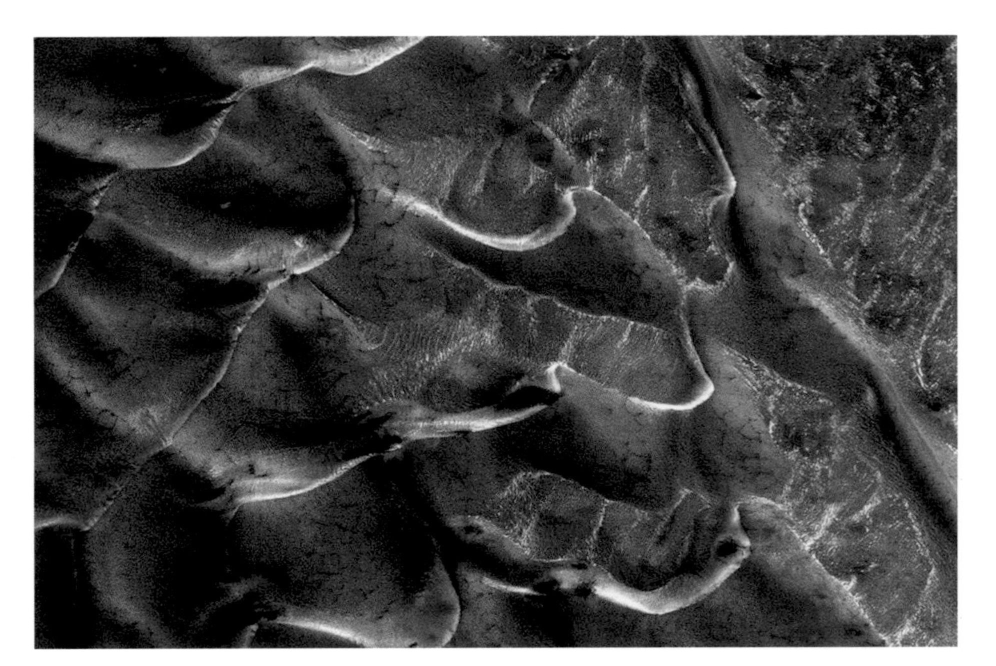

Sand Dunes

A field of sand dunes occupies this frosty 5-kilometer diameter crater in the high-latitudes of the northern plains of Mars. Some dunes have separated from the main field and appear to be climbing up the crater slope along a gully-like form. The surface of the main dune field is characterized by a series of dark-toned polygonal patterns. These may be the result of seasonal frost processes. Several of the steeper dune slopes, pointing in the downwind direction, host narrow furrows suggesting the start of gully formation.

[Photo and caption credit: NASA/JPL-Caltech/University of Arizona]

Brain Terrain

You are staring at one of the unsolved mysteries on Mars. This surface texture of interconnected ridges and troughs, referred to as "brain terrain" is found throughout the mid-latitude regions of Mars. (This image is in an area of Mars called Protonilus Mensae.) This bizarrely textured terrain may be directly related to the water ice that lies beneath the surface. One hypothesis is that when the buried water ice sublimates (changes from a solid to a gas), it forms the troughs in the ice. The formation of these features might be an active process that is slowly occurring since HiRISE has yet to detect significant changes in these terrains.

[Photo and caption credit: NASA/JPL-Caltech/University of Arizona]

Starburst Spider

Mars' seasonal cap of carbon dioxide ice has eroded many beautiful terrains as it sublimates every spring. In the region where the HiRISE on NASA's Mars Reconnaissance Orbiter took this image, we see troughs that form a starburst pattern. In other areas these radial troughs have been referred to as spiders, simply because of their shape. In this region the pattern looks more dendritic as channels branch out numerous times as they get further from the center. The troughs are believed to be formed by gas flowing beneath the seasonal ice to openings where the gas escapes, carrying along dust from the surface below. The dust falls to the surface of the ice in fan-shaped deposits.

[Photo and caption credit: NASA/JPL-Caltech/University of Arizona]

Concepts for Mars Sample Return

When plans began for the Mars 2020 mission and its Perseverance rover (a collaboration between NASA and ESA), the spacecraft's design included the ability to collect soil samples for storage and retrieval in the future. If all goes as planned, the NASA-led sample retrieval lander would launch to the Red Planet in 2026, landing near Perseverance rover in the Jezero Crater in 2028. The lander would retrieve the samples and then place the canister in the return rocket for launch into orbit around Mars. However, there is also the possibility that Perseverance would be able to deliver the samples to the return rocket. NASA's sample return rocket would then lift off to rendezvous with the ESA Earth return orbiter and, in doing so, would be the first vehicle to have lifted off from another planet.

The ESA Earth return orbiter is also planned for launch in 2026. After retrieving Mars soil samples from Perseverance, the NASA's sample return spacecraft would rendezvous with the ESA vehicle and transfer the soil canister for the trip back to Earth. This would include a spacecraft designed to enter the Earth's atmosphere and safely land at a designated site in the United States. As was the case with soil samples from the Moon, scientists will have an opportunity to learn more about another planet and the possibility of past or present microbiological life on Mars. Learning more about the solar system's past may give a better understanding for the future.

Above: This illustration shows a concept for a set of future robots working together to ferry back samples from the surface of Mars, collected by NASA's Mars Perseverance rover.

[Picture and caption credit: NASA/ESA/JPL-Caltech]

6: Interplanetary Explorations

Interplanetary exploration—is it mankind's next logical step or are we just burning through earth's resources at breakneck speed? From my humble perspective, its probably a bit of both. As a species grounded in curiosity and materialism, humankind seems more and more like a two-headed creature with the ability to sprint in both directions simultaneously. That said, with the recent launch of the James Webb telescope—the world's most powerful optical instrument—we'll be able to receive images going back over four billion years. The understanding of where we've been and where we've come from have a direct correlation to where we are and where we're going. It's hard to grasp the impact of how this new understanding about our solar system will help or hinder our existence, but we've opened the next chapter using unprecedented technology and a fierce appetite to explore. So let's wait and see what happens.

Pioneer

In August 1958, two months before the establishment of NASA, the Air Force Ballistic Missile Division attempted to place a small spacecraft in orbit around the Moon. The first attempt to reach past Earth ended in the failure of the Thor-Able launch vehicle, but many successes have followed.

Shortly after NASA's formation on 1 October 1958, the agency launched the first successful mission to explore beyond Earth. On 3 March 1959, Pioneer 4 lifted off aboard the Army Ballistic Missile Agency's Juno II rocket and became the first American spacecraft to escape Earth's gravity. The small probe weighed 13 pounds and measured 1.67 feet in length and 9 inches in diameter. It passed within 37,000 miles of the Moon, providing valuable data on radiation and tracking objects in space.

NASA launched Pioneer 5 on 11 March 1960 to study the area in space between Earth and the planet Venus. The spacecraft was placed in a heliocentric orbit circling the Sun.

Bottom left: Early 1958 Pioneer lunar orbiter. This mission was the first of two US Air Force (USAF) launches to the Moon and was the first attempted deep space launch by any nation.

[Photo and caption credit: NASA/JPL]

Bottom right: NASA's Pioneer 5 spacecraft, nicknamed the "Paddle-Wheel Satellite," before launch.

[Photo credit: NASA]

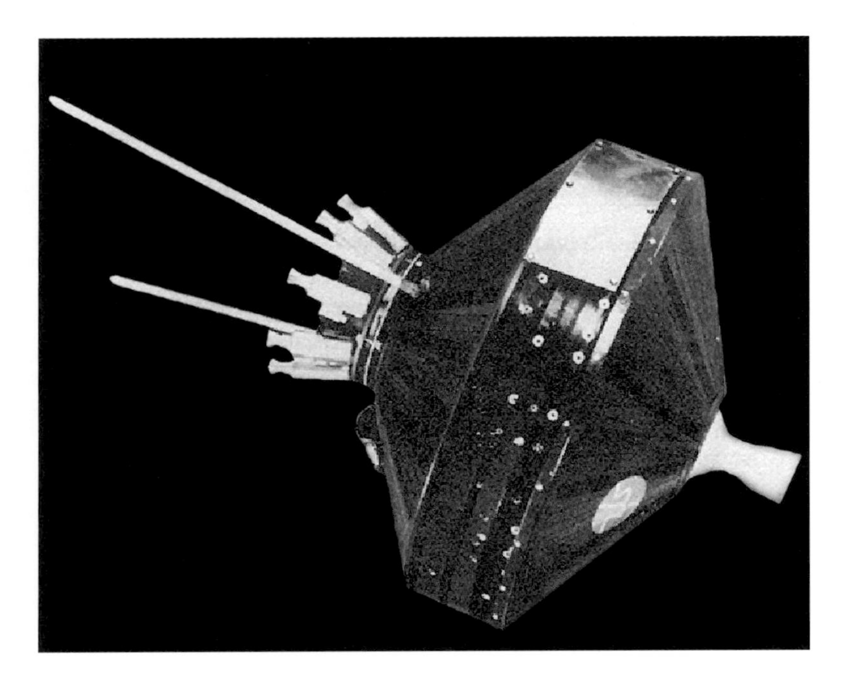

It successfully measured particles from solar flares and cosmic radiation in interplanetary space. Pioneer 5's achievements include the first use of Telebit, a digital telemetry system that was a significant improvement over previous systems.

Above: Our solar system showing the planets, along with their respective moons. From right to left: Sun, Mercury, Venus, Earth, Mars, Jupiter, Saturn, Uranus, and Neptune.

Ranger 3

NASA's Ranger Program was a series of uncrewed spacecraft launched to obtain the first close-up images of the surface of the Moon. While the first six suffered various failures, one of them, Ranger 3, did successfully collect data on interplanetary gamma rays.

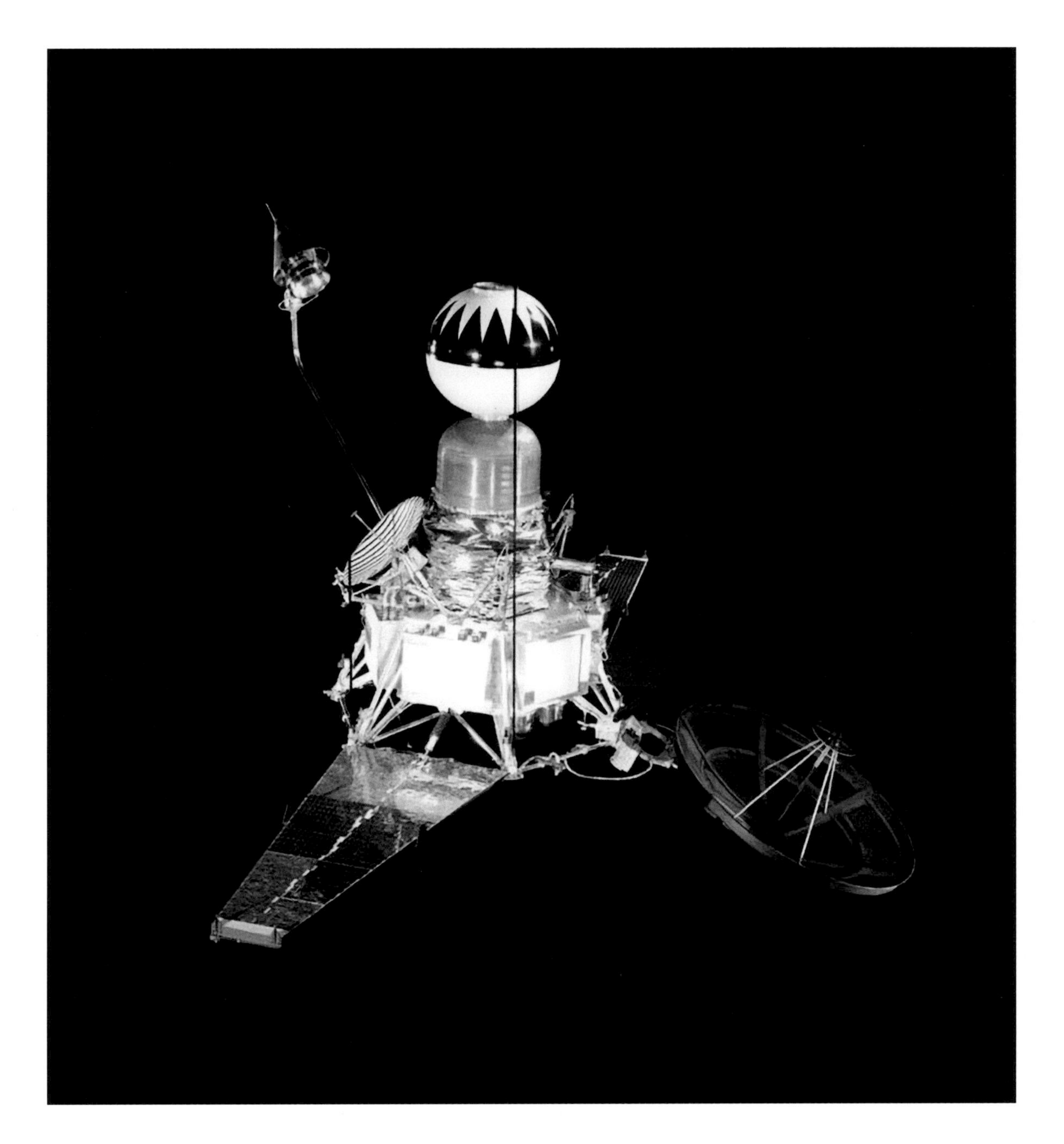

Right: NASA's Ranger spacecraft.

[Photo credit: NASA]

Ranger 7

Ranger 7 was the first to transmit detailed pictures in the final 17 minutes before impact with the Moon on 31 July 1964. Rangers 8 and 9 followed in February and March 1965, each sending back photographs that were valuable in the planning for landing Apollo astronauts on the lunar surface.

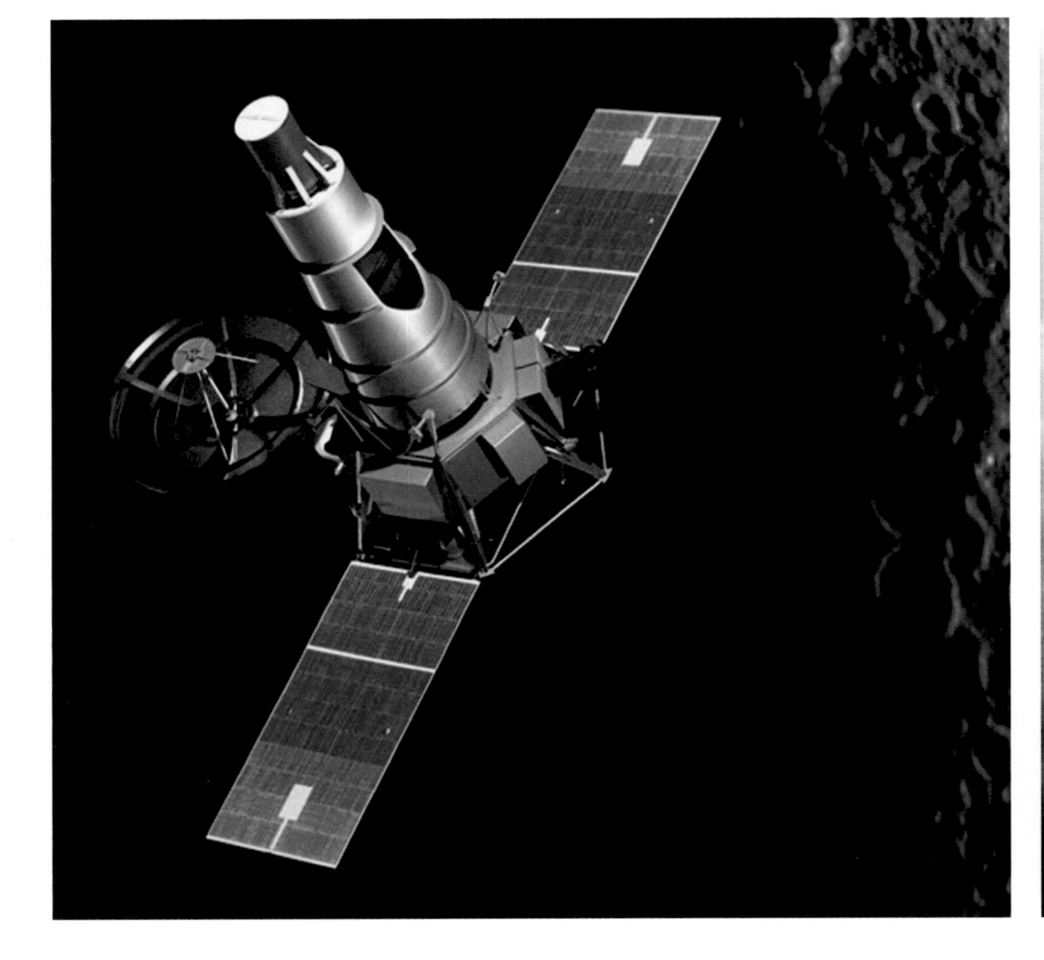

Above: An artist's concept of NASA's Ranger spacecraft approaching the Moon.

[Picture and caption credit: NASA/JPL-Caltech]

Above: One of 4,316 images Ranger 7 sent back before impact.

[Photo and caption credit: NASA/JPL-Caltech]

Mariner Program

The first spacecraft encounter with another planet in the solar system was Mariner 2, similar in design to the Ranger probes which headed towards the Moon. During its trip to Venus, Mariner 2 completed measurements of the solar wind, a perpetual stream of particles ejected from the Sun. During its fly-by of Venus on 14 December 1962, the spacecraft's instruments determined that the surface temperatures were very hot and the atmosphere was so thick as to be opaque. The mission of the Mariner 5 probe was to perform radio-occultation experiments to further determine properties of the thick Venusian atmosphere. On its way, the spacecraft made the closest pass to the Sun at the time. As Mariner 5 approached

Below: NASA's Mariner 5 spacecraft.
[Photo credit: NASA]

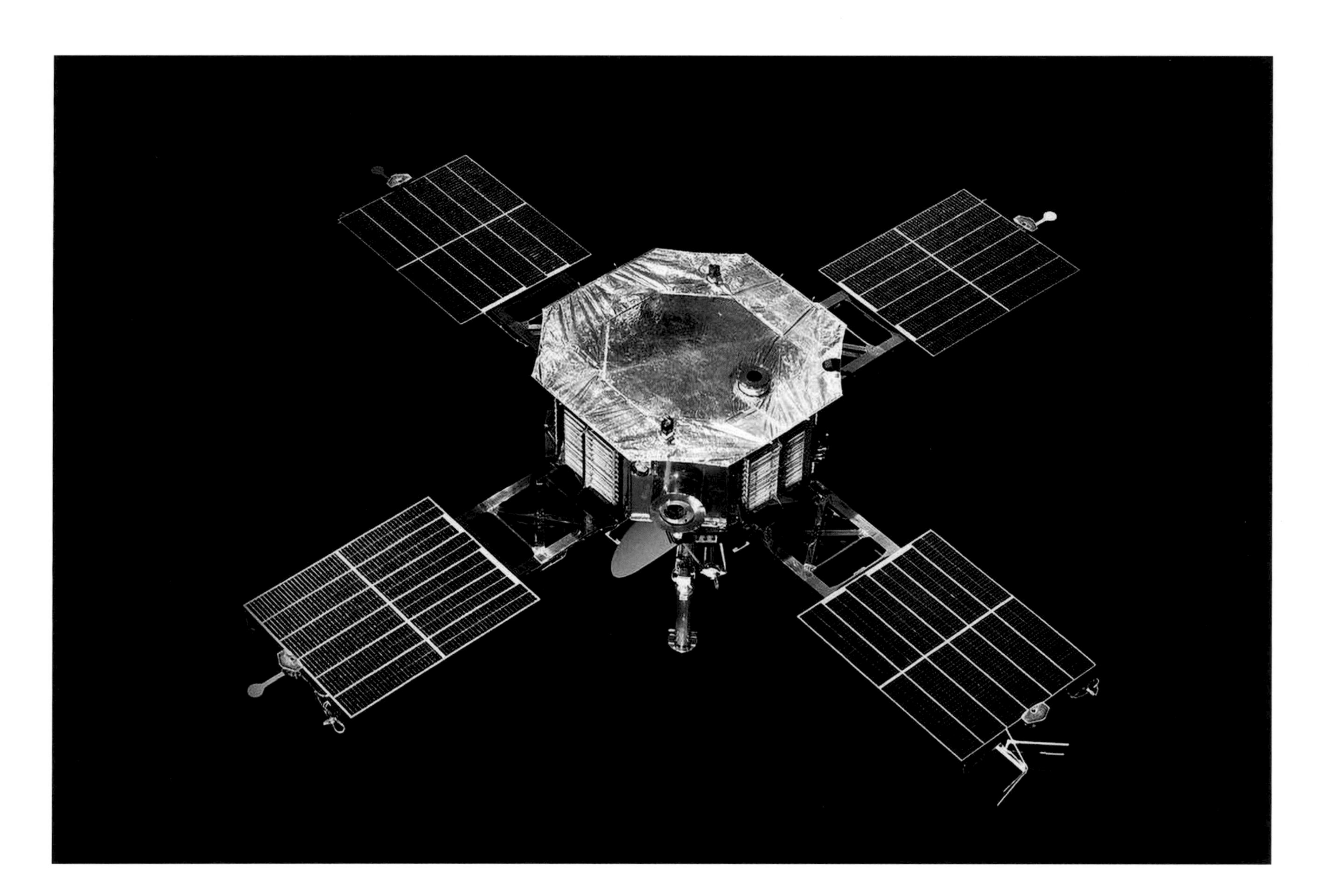

Venus, its instruments found that the planet does not have a magnetic field and the dense atmosphere deflects the solar wind.

At the same time, Mariner 4 was the first spacecraft to fly past Mars, returning photographs in July 1965. As a follow-on, in August 1969, NASA's Mariner 7 was programmed to focus on the Mars' south polar region. The spacecraft also returned images of the Martian moon, Phobos, indicating that it was irregularly shaped.

Left: This photo shows the weight-and-center-of-gravity test conducted on flight spacecraft M69-2, about three months before it was shipped to the Air Force Eastern Test Range (AFETR) at Cape Canaveral. [Photo and caption credit: NASA/JPL-Caltech]

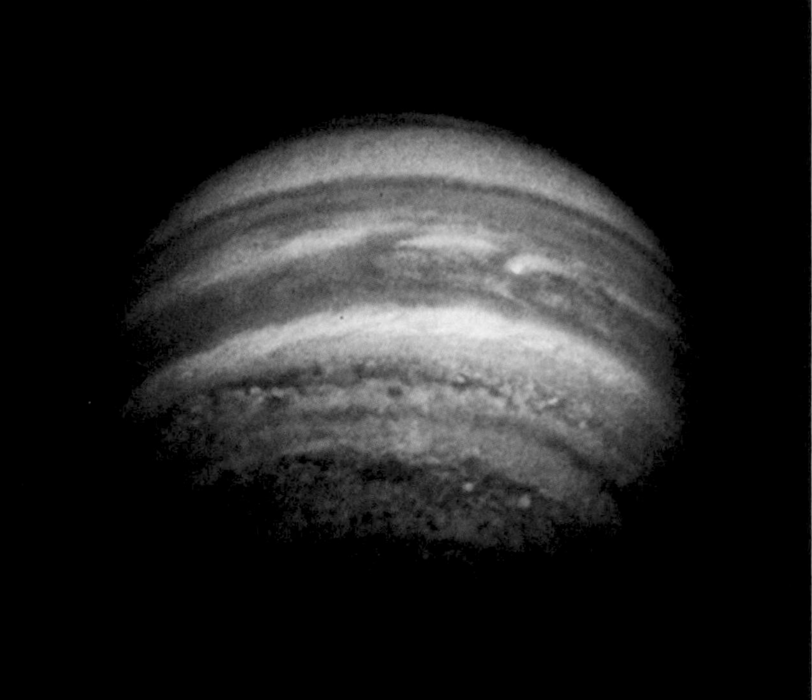

Pioneers 10 and 11

Launching on 2 March 1972, NASA's Pioneer 10 spacecraft was the first to explore the planet Jupiter, where it began transmitting photographs 20 months later. During more than 30 years of operation, Pioneer 10's instruments studied the area near Jupiter and beyond, including measurements of the solar wind, cosmic rays, and farthest areas of the solar system. Launched on 5 April 1973, Pioneer 11 photographed Jupiter and returned data similar to that of its sister spacecraft. In addition to taking images of Jupiter's polar areas, Pioneer 11 recorded about 200 photographs of the planet's moons. This second deep-space probe continued on to Saturn. Its closest approach with the ringed planet was on 3 December 1974, with its instruments confirming that Saturn has a magnetic field.

Viking Landers

In August and September 1975, two Viking spacecraft were launched three weeks apart for missions to Mars. The twin orbiters included spacecraft designed to soft land on the surface. Both Vikings were intended to seek evidence of life on Mars in the present or the past. After successfully touching down on the surface, the Viking 1 lander began transmitting high-quality color photographs of the nearby area. The high-resolution cameras also took panoramic images, including the gently rolling plains close to the landing site. Both Vikings 1 and 2 collected science data and conducted geological investigations.

Opposite top: An artist's concept of NASA's Pioneer 11 spacecraft.

[Photo and caption credit: NASA]

Opposite bottom, left: NASA's Pioneer 11's path through Saturn's outer rings took it within 13,000 miles (21,000 kilometers) of the planet.

[Photo and caption credit: NASA/Ames]

Opposite bottom, right: Jupiter as seen from above its north pole by Pioneer 11 in 1974.

[Photo and caption credit: NASA]

Below: This is the first photograph ever taken on the surface of Mars. It was obtained by NASA's Viking 1 minutes after the spacecraft landed on 20 July 1976.

[Photo and caption credit: NASA/JPL-Caltech]

Voyager 1 and 2

In 1977, NASA launched two spacecraft called, "Voyager," to the solar system's outer planets. In the case of Voyager 2, during its 12 years' journey it flew past four planets on a mission known as the "Grand Tour." Voyager 2 was launched first, lifting off on 20 August 1977, 16 days before its fellow traveler, Voyager 1. Voyager 2 proceeded on a trajectory taking longer to reach Jupiter and Saturn, but using gravity assists from each to continue farther to Uranus and Neptune.

On 5 September 1977, Voyager 1 launched on its path to Jupiter, arriving first in January 1979. Its closest encounter with the planet was on 5 March 1979 at about 217,000 miles. The spacecraft was able to take high-resolution images of the planet's rings and moons, while measuring its magnetic field and radiation belt.

Voyager 2 made its closest approach to Jupiter in July 1979, traveling within 350,000 miles. The spacecraft determined that the planet's large red spot was a complex storm.

First to arrive at Saturn was Voyager 1, coming closest at 77,000 miles on 12 November 1980. Cameras on the spacecraft found complex structures in the rings of Saturn looking like layers of rock. Additionally, sensing instruments examined Saturn's atmosphere and that of its giant moon, Titan.

Right: Voyager 2 took this image as it approached the planet Uranus on 14 January 1986. The planet's hazy bluish color is due to the methane in its atmosphere, which absorbs red wavelengths of light.

[Photo and caption credit: NASA/JPL-Caltech]

The closest approach to Saturn for Voyager 2 was on 26 August 1981. While passing Saturn, the spacecraft surveyed its atmosphere, collecting information on its density and temperature.

More than eight years into Voyager 2's "Grand Tour," the spacecraft flew past Uranus finding it to be a cold planet with temperatures at 353 degrees below zero Fahrenheit. During its closest approach on 24 January 1986, Voyager 2 traveled as close as 50,600 miles of the planet's atmosphere. In doing so, Voyager 2 collected data exposing more rings and 11 additional moons.

NASA's Voyager 2 made its closest flyby of Neptune on 25 August 1989. The planet was found to have blue hues of teal and cobalt, signifying the presence of methane in the atmosphere. Neptune has what was termed a "great dark spot." Like Jupiter's red spot, it is a large storm on the planet.

On 25 August 2012, Voyager 1 became the first human-made object to cross the heliopause at the outer edge of the solar system. Both Voyagers 1 and 2 are now on extended missions investigating interstellar space.

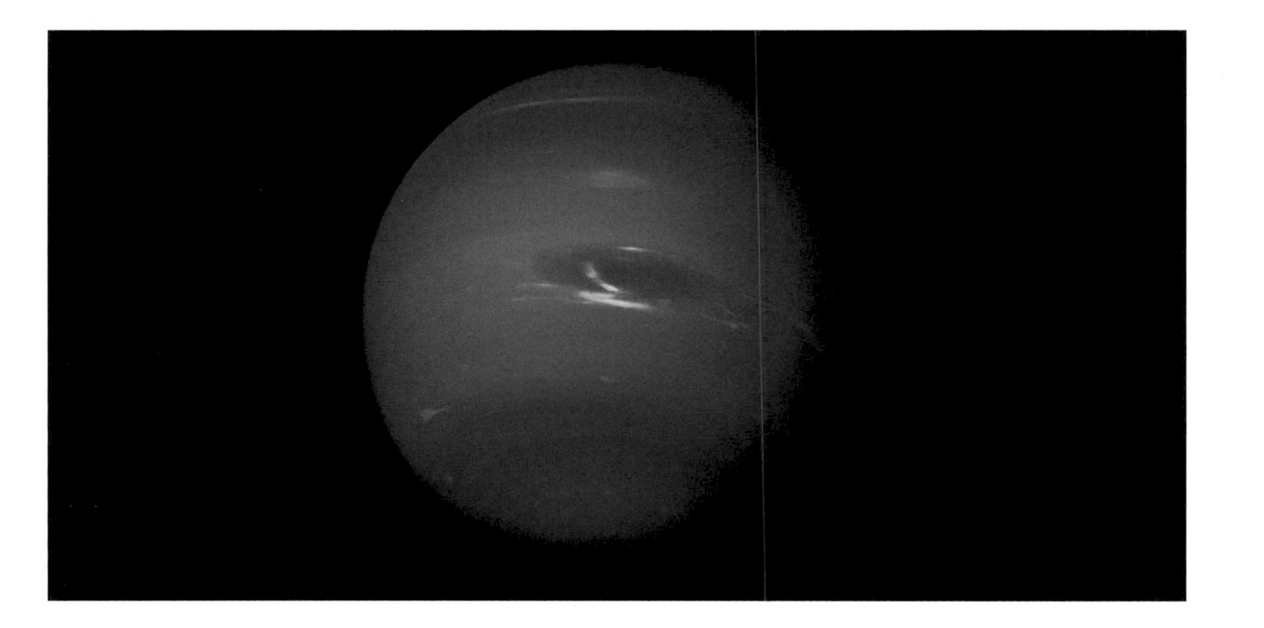

Left: This picture of Neptune was taken by Voyager 2 less than five days before the probe's closest approach of the planet on 25 August 1989. The picture shows the "great dark spot"—a storm in Neptune's atmosphere—and the bright, light-blue smudge of clouds that accompanies the storm.

[Photo and caption credit: NASA/JPL-Caltech]

Pioneer Venus

The Pioneer Venus orbiter was launched in 20 May 1978 to study clouds and infrared emissions in the planet's atmosphere, as well as using radar to map the surface's topography and characteristics. It operated from December 1978 until October 1992. A similar spacecraft, the Pioneer Venus multi-probe, also known as Pioneer Venus 2, lifted off on 8 August 1978. The goal was to release three probes into the planet's dense atmosphere in different locations on 9 December that same year. The instruments determined that convection does not exist in Venus's atmosphere and, below a layer of haze at a distance of about 19 miles above the surface, the atmosphere is almost clear.

Right: An artist's illustration of NASA's Pioneer Venus 2 approaching Venus.

[Picture and caption credit: NASA/ Paul Hudson]

SOHO

The Solar and Heliospheric Observatory, or SOHO, spacecraft was a joint effort between NASA and the European Space Agency to study the Sun. Launched on 2 December 1995, with 12 instruments to conduct research into the Sun's atmosphere and the solar wind, SOHO has provided scientists with a better understanding of the Sun's internal structure and dynamics.

This spacecraft was designed for a two-year mission but its operation has been prolonged to the end of 2025, allowing observation of two 11-year solar cycles. As a result, researchers are gaining insights into how the Sun affects life on Earth, such as forecasting possibly perilous solar storms.

Below: An artist's concept of the ESA-NASA SOHO spacecraft.

[Picture and caption credit: NASA]

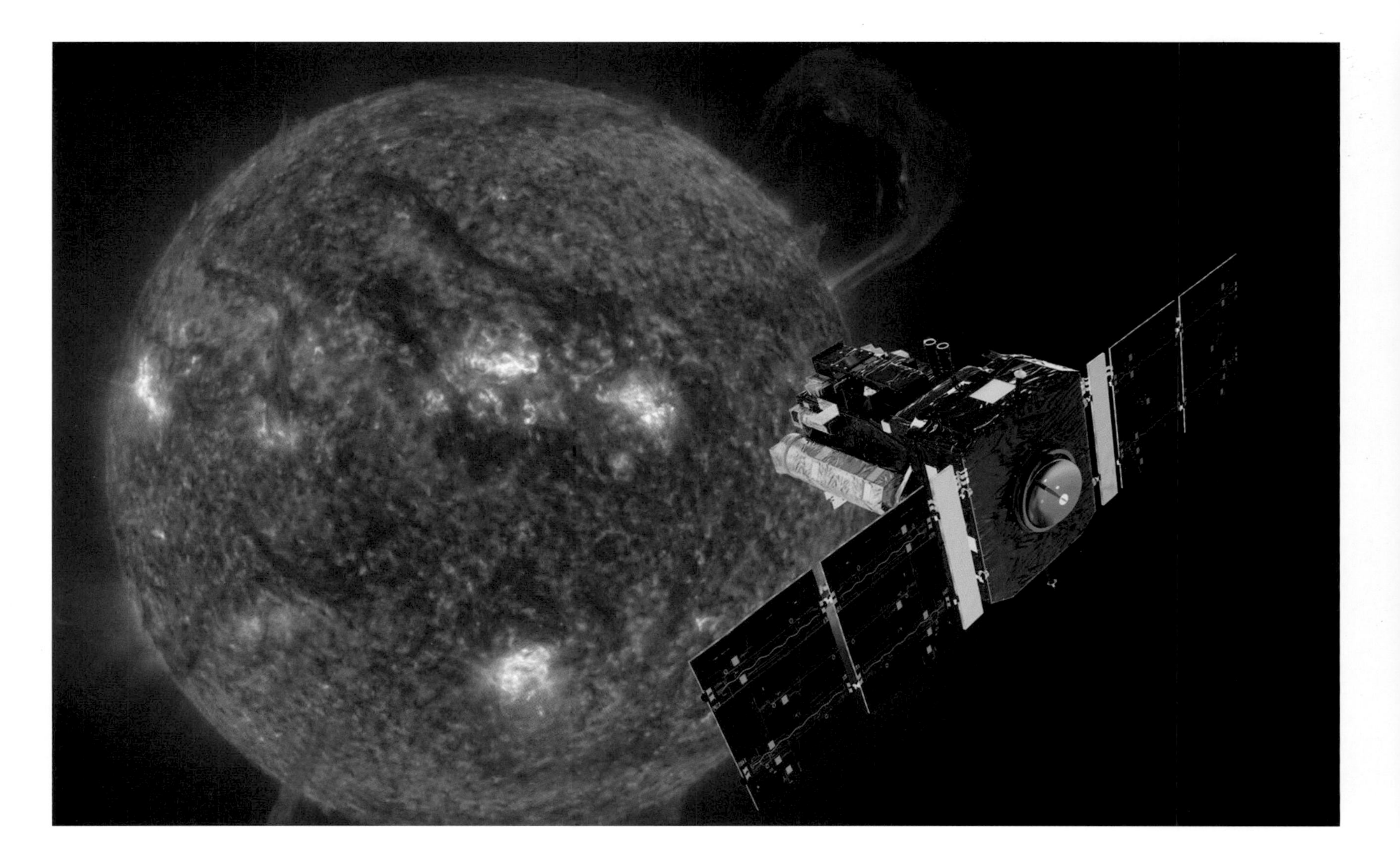

“Invention is the most important product of man’s creative brain. The ultimate purpose is the complete mastery of mind over the material world, the harnessing of human nature to human needs.”

Nikola Tesla, inventor, electrical and mechanical engineer, and futurist

Left: The Pioneer Venus Multiprobe is being lifted on a crane in Hangar AO on Cape Canaveral Air Station. Technicians are doing their last inspection of the 2,000 pound spacecraft before it is enclosed in its protective nose fairing.

Deep Space 1

In October 1998, NASA launched a spacecraft called Deep Space 1 to test new, advanced technology for future flights to interplanetary destinations and farther into deep space. The goal was to fly near both an asteroid and a comet. The innovations included ion propulsion, autonomous optical navigation, and an instrument that was both a small camera and an imaging spectrometer. Deep Space 1 completed its closest asteroid fly-by to date in July 1999, passing the asteroid 9969 Braille. In September 2001, the probe completed an encounter with the comet Borrelly. Both fly-bys provided some of the best images and data from an asteroid or a comet.

Right: An artist's rendering of NASA's Deep Space 1.
[Picture and caption credit: JPL/NASA]

Deep Impact

Deep Impact was a NASA mission to learn more about the nature and interior composition of comets. The spacecraft was launched on 12 January 2005 to rendezvous with, and impact, the comet Tempel 1. On 4 July 2005, the probe's impactor was successful in hitting the comet's nucleus, forming a crater. In addition to transmitting more than 500,000 photographs of celestial objects, Deep Impact uncovered much about the interior makeup of a comet.

Below: An artist's impression of the Deep Impact spacecraft. [Photo and caption credit: NASA]

MESSENGER to Mercury

Mercury is the closest planet to the Sun and its proximity to the hotter environment at the center of the solar system presents inherent challenges. NASA's first expedition to the small planet was Mariner 10 making three brief fly-bys during 1974 and 1975. The first effort to orbit a spacecraft around Mercury was NASA's MESSENGER (MErcury Surface, Space ENvironment, GEochemistry and Ranging) mission. The spacecraft was designed and operated for NASA by the Johns Hopkins University Applied Physics Laboratory in Maryland.

Launched on 3 August 2004, MESSENGER set out on a 4.9-billion-mile trip, requiring over six and a half years to reach its destination. In doing so, the spacecraft performed several gravity-assisted fly-bys: once of Earth, twice of Venus, and three times of Mercury, allowing the use of

Right: Illustration of MESSENGER in orbit around Mercury.

[Picture and caption credit: NASA]

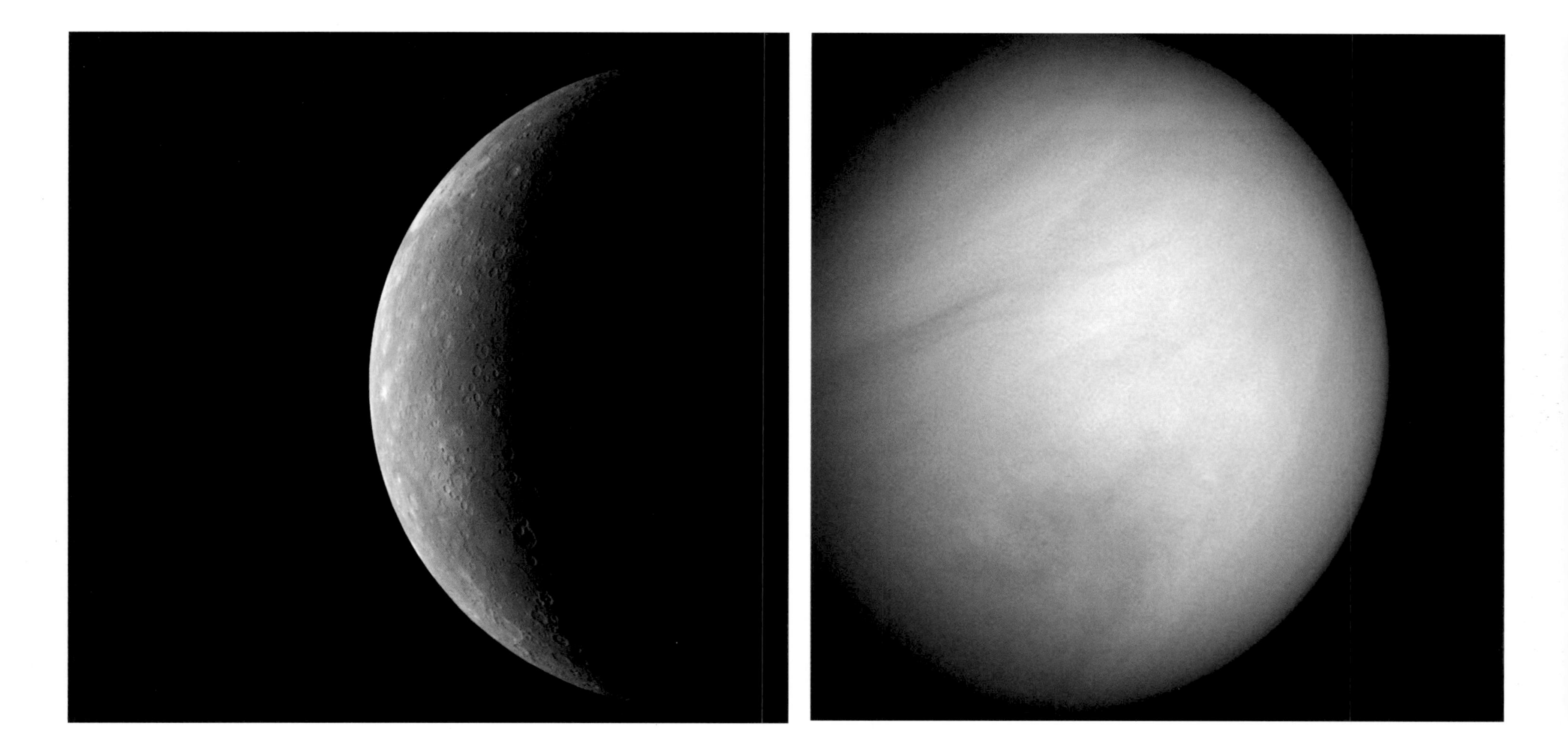

minimal fuel. MESSENGER became the first to orbit the planet Mercury on 18 March 2011. Upon arrival, the spacecraft's instruments found evidence of water ice and organic compounds at Mercury's north pole that includes regions that are permanently shaded from the Sun. MESSENGER transmitted extensive science data along with over 700 photographs of comets. With its systems continuing to work well, there were two mission extensions, with MESSENGER sending its final photograph on 30 April 2015.

Above left: Photograph of Mercury during a flyby in January 2008.

Above right: Photograph of Venus during a flyby in July 2006.

[Photo and caption credits: NASA]

"Once you can accept
the universe as matter
expanding into nothing that
is something, wearing stripes
with plaid comes easy."

Albert Einstein

Above: MESSENGER's first photograph taken from orbit around Mercury.

Above: Phctograph of Crater Stevenson with perpendicular crater chains.

Above: Sunlit side of planet Mercury.

Above: MESSENGER's last photograph.

[Photo and caption credits: NASA/John Hopkins University Applied Physics Laboratory/Carnegie Institution of Washington]

Above: An artist's rendition of ESA's Venus Express spacecraft in orbit around Venus.

[Picture and caption credit: ESA - D. Ducros]

Venus Express

The first European spacecraft to orbit the planet Venus was the European Space Agency's (ESA's) Venus Express. It was launched on 9 November 2005 by a Russian Soyuz rocket, lifting off from the Baikonur Cosmodrome in Kazakhstan. Following a five-month trip, Venus Express entered an elliptical-polar orbit around Venus on 11 April 2006.

Venus Express was originally scheduled to last about 500 days, however, the life of the spacecraft was extended five times through 2015. During its nine-year mission, Venus Express developed a temperature map of the southern hemisphere of Venus. Additional discoveries pointed to the possibility of oceans on the Venusian surface and more lightning on Venus than on Earth.

New Horizons

NASA's New Horizons spacecraft launched atop an Atlas V rocket on 19 January 2006 to explore Pluto, the farthest planet from the Sun. It is also studying the area at the outer edge of the solar system known as the Kuiper Belt.

There is no universal agreement among astronomers as to whether Pluto should be referred to as a planet. After the launch of New Horizons in 2006, the International Astronomical Union redefined Pluto as a "dwarf planet." NASA's principal investigator for New Horizons, Alan Stern, associate vice president for Research and Development at the Southwest Research Institute in Colorado, disagrees with that definition and still describes Pluto as a planet.

"We're just learning that a lot of planets are small planets and we didn't know that before," he said, just before New Horizon's fly-by of Pluto in 2015. "Fact is, in planetary science, objects such as Pluto and the other dwarf planets in the Kuiper Belt are considered planets and called planets in everyday discourse in scientific meetings."

New Horizons made its closest approach to Pluto on 14 July 2015 following a nine-year trip of more than 3.6 billion miles. The spacecraft transmitted the first-ever close-up images and scientific observations of the distant planet. The up-close observations allowed astronomers to confirm the size of Pluto as about 1,470 miles in diameter, and its largest moon, Charon, as about 750 miles in diameter. In late 2018, New Horizons began passing through the Kuiper Belt that includes an estimated 100,000-plus objects, some of which have diameters no larger than 62 miles.

Below: NASA's New Horizons unmanned spacecraft approaches dwarf planet Pluto and its moon Charon. [Picture and caption credit: Alamy Stock Photo]

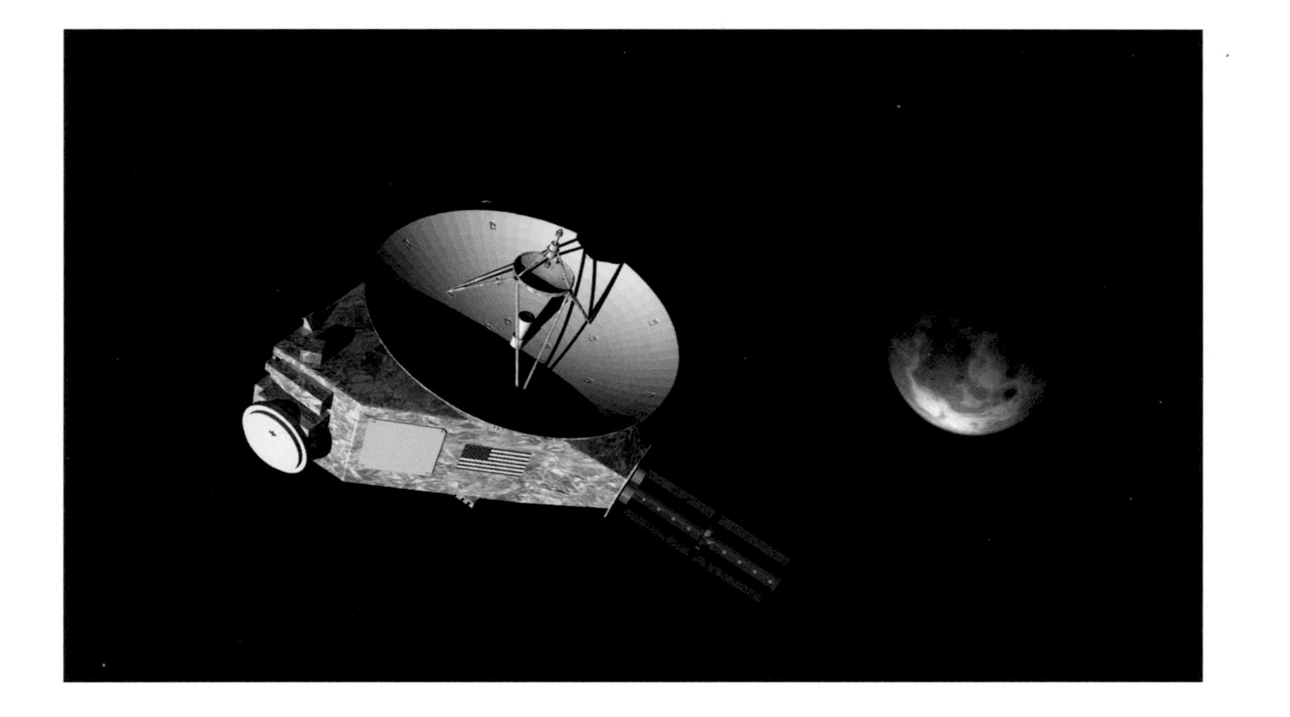

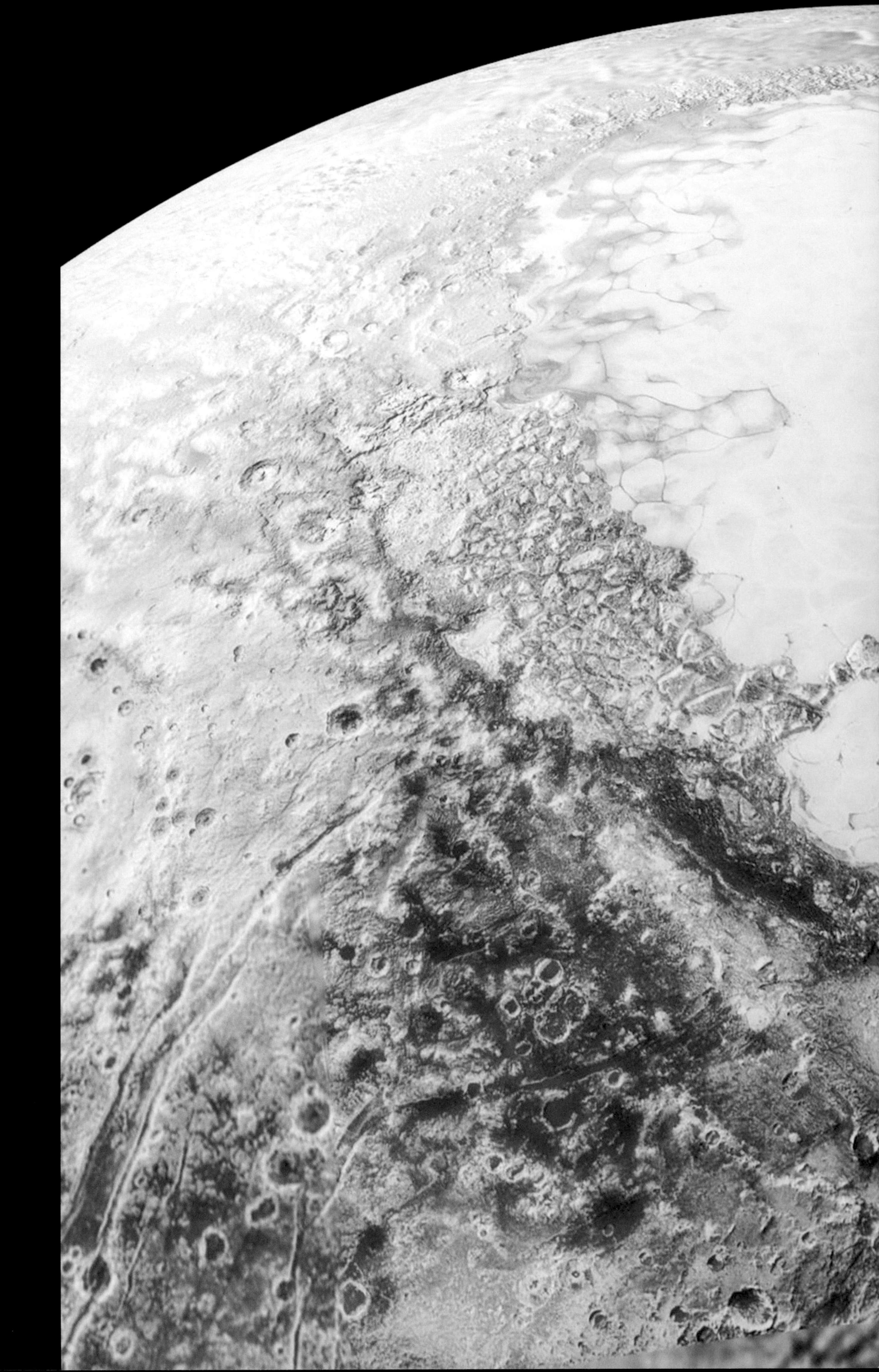

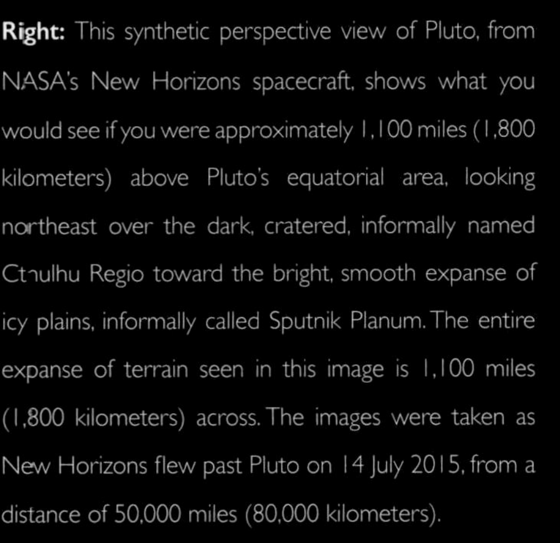

Right: This synthetic perspective view of Pluto, from NASA's New Horizons spacecraft, shows what you would see if you were approximately 1,100 miles (1,800 kilometers) above Pluto's equatorial area, looking northeast over the dark, cratered, informally named Cthulhu Regio toward the bright, smooth expanse of icy plains, informally called Sputnik Planum. The entire expanse of terrain seen in this image is 1,100 miles (1,800 kilometers) across. The images were taken as New Horizons flew past Pluto on 14 July 2015, from a distance of 50,000 miles (80,000 kilometers).

[Photo and caption credit: NASA/Johns Hopkins University Applied Physics Laboratory/ Southwest Research Institute]

WISE and NEOWISE

NASA's Wide-field Infrared Survey Explorer, or WISE, is a space telescope that records astronomical imagery in infrared wavelengths. It was launched on 14 December 2009. WISE was successful in discovering thousands of small distant planets and star clusters beyond the solar system. After being set into a sleep mode in February 2011, the telescope was re-activated in August 2013 and renamed NEOWISE (Near-Earth Object Wide-field Infrared Survey Explorer). The spacecraft is now looking to discover any unknown objects near Earth and learn more about known asteroids, there being as many as 2,000.

Below: An artist's concept of NASA's WISE/NEOWISE spacecraft.

[Picture and caption credit: NASA/JPL-Caltech]

LADEE

On 7 September 2013, NASA launched an 844-pound spacecraft to orbit the Moon to study dust in the area, as well as its thin atmosphere known as the lunar exosphere. Called LADEE, for Lunar Atmosphere and Dust Environment Explorer, the probe's instrumentation helped scientists better understand the composition and density of the Moon's exosphere, which is similar to that of the planet Mercury and some of the moons of Jupiter. The mission concluded on 18 April 2014.

Below: An artist's concept of LADEE at the Moon. [Picture and caption credit: NASA Ames / Dana Berry]

Gaia

Like WISE and NEOWISE, the Global Astrometric Interferometer for Astrophysics, or GAIA, is a space-based observatory. The spacecraft was launched on 19 December 2013 by the European Space Agency. The objective is to develop a precise, three-dimensional diagram of the stars in the Milky Way galaxy. GAIA will do this by charting about one percent of the Milky Way's 100 billion stars and other objects such as planets, comets, and asteroids. This will be achieved by focusing on objects an average of 70 times during a five-year period, recording brightness and the position of each object. Plans call for GAIA to continue in operation through 2025.

Below: An artist's concept of the Gaia spacecraft. [Picture credit: ESA/ATG medialab; background: ESO/S. Brunier]

BepiColombo

After NASA's Mariner 10 and MESSENGER missions to the planet Mercury, ESA and the Japan Aerospace Exploration Agency (JAXA) joined together for BepiColombo. Two spacecraft were launched together to Mercury on 20 October 2018 on an Ariane 5 rocket, lifting off from Guiana Space Centre in Kourou, French Guiana. Like MESSENGER, BepiColombo is making several fly-bys of Earth, Venus, and Mercury before entering its planned orbit around the planet closest to the Sun, targeted for 5 December 2025.

The main spacecraft, the Mercury Planetary Orbiter, was produced by ESA. Its instruments are designed to investigate the surface and internal composition of the planet. The Mercury Magnetospheric Orbiter was provided by JAXA. This portion of BepiColombo will examine the region of space around Mercury known as the magnetosphere that is dominated by its magnetic field.

Below left: An artist's impression of the ESA–JAXA BepiColombo spacecraft.

[Picture credit: ESA/ATG medialab]

Below right: The BepiColombo mission captured this view of Mercury on 1 October 2021 as the spacecraft flew past the planet for a gravity assist manoeuvre.

[Photo and caption credit: ESA/BepiColombo/MTM]

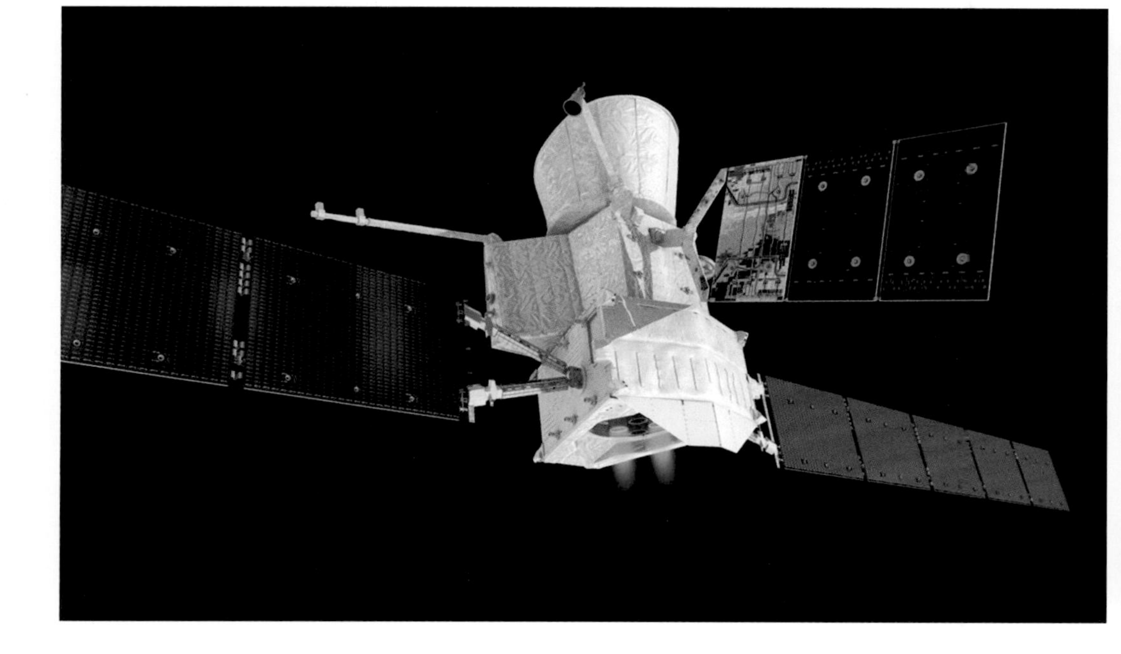

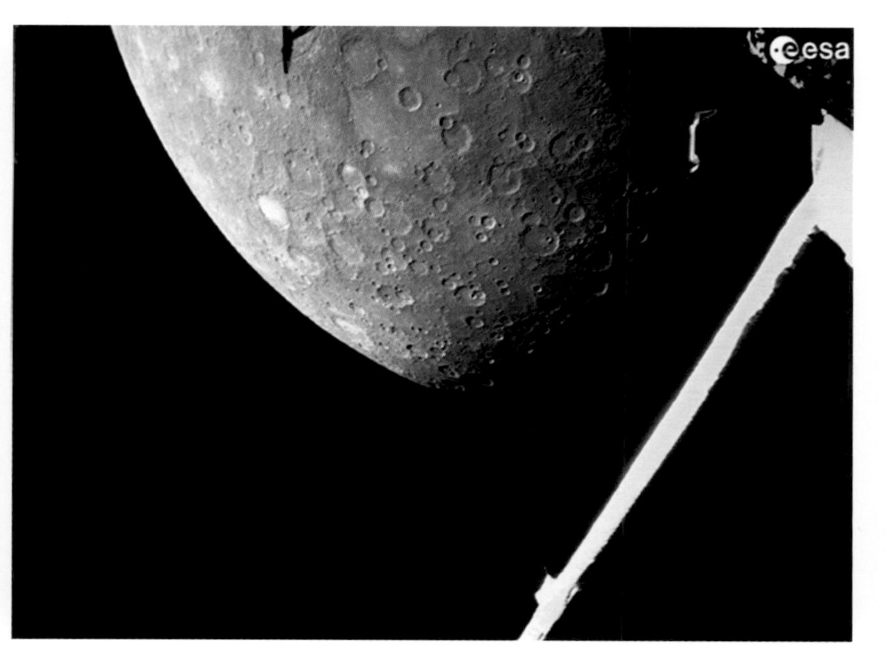

Asteroids, Comets, and Meteors

As NASA has studied objects in the solar system, the focus has been on the Moon, the Sun, and the planets. In recent years, spacecraft have been launched to better understand other objects such as asteroids, comets, and meteors that help make up our "neighborhood" in space.

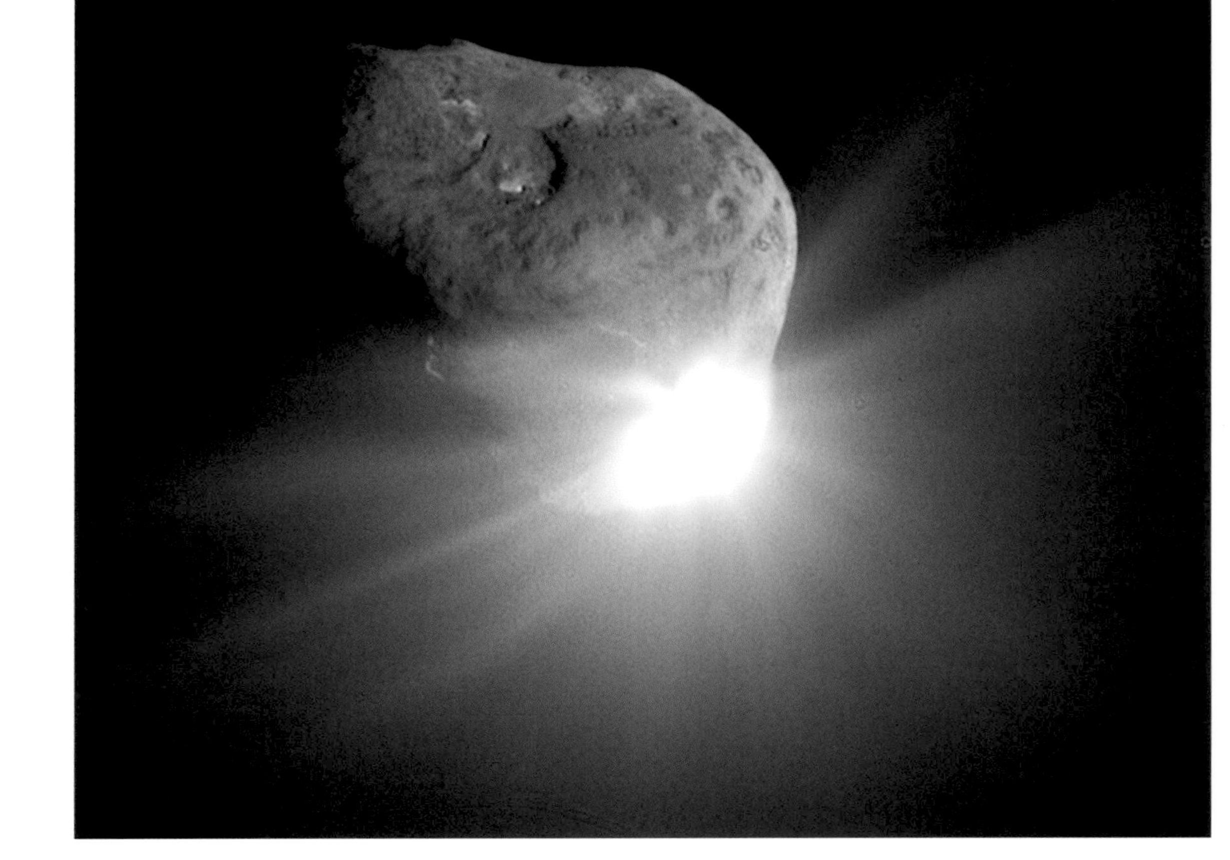

Right: This spectacular image of comet Tempel I was taken 67 seconds after it obliterated Deep Impact's impactor spacecraft. The image was taken by the high-resolution camera on the mission's fly-by craft. Scattered light from the collision saturated the camera's detector, creating the bright splash seen here. Linear spokes of light radiate away from the impact site, while reflected sunlight illuminates most of the comet surface. The image reveals topographic features, including ridges, scalloped edges, and possibly impact craters formed long ago.
[Photo and caption credit: NASA/JPL-Caltech/UMD]

Comets

Comets are leftovers of the origins of the solar system, sometimes called, "dirty snowballs." Astronomer Gerard Kuiper published a theory in 1951 stating that he believed comets are a disc-like belt of icy bodies beyond the planet Neptune where dark comets orbit the Sun near Pluto. Comets are actually made up mostly of frozen gases, rock, and dust. As a comet's trajectory approaches the Sun, it heats up developing an atmosphere and emitting dust and gases that become a large glowing head or nucleus that may be only a few miles across. Comets also develop a tail that can stretch for millions of miles.

Left: A section of the smaller Comet 67P/Churyumov–Gerasimenko's two lobes, as seen through Rosetta's narrow-angle camera from a distance of about 8 km to the surface on 14 October 2014. The resolution is 15 cm/pixel. The image is featured on the cover of the 23 January 2015 issue of the journal *Science*.

[Photo and caption credit: ESA/Rosetta/MPS for OSIRIS Team]

Meteoroids, Meteors, and Meteorites

Below: Meteor Crater (also known as Barringer Crater) on Earth is only 50,000 years old. Even so, it's unusually well preserved in the arid climate of the Colorado Plateau. Meteor Crater formed from the impact of an iron-nickel asteroid about 46 meters (150 feet) across. Most of the asteroid melted or vaporized on impact. The collision initially formed a crater over 1,200 meters (4,000) feet across and 210 meters (700 feet) deep. Subsequent erosion has partially filled the crater, which is now only 150 meters (550 feet) deep. Layers of exposed limestone and sandstone are visible just beneath the crater rim, as are large stone blocks excavated by the impact.

[Photo and caption credit: NASA]

These are relatively the same and are the streaking lights sometimes seen as "shooting stars." **Meteoroids** are objects in space ranging from grains of dust to minor asteroids. The difference may depend on where they are. When "space rocks" enter the Earth's atmosphere, they begin to burn up, creating a fireball defined as a **meteor**. If portions of the rock survive and land, it is referred to as a **meteorite**.

Left: Designated Northwest Africa (NWA) 7034, and nicknamed "Black Beauty," this Martian meteorite weighs approximately 11 ounces (320 grams).

[Photo and caption credit: NASA]

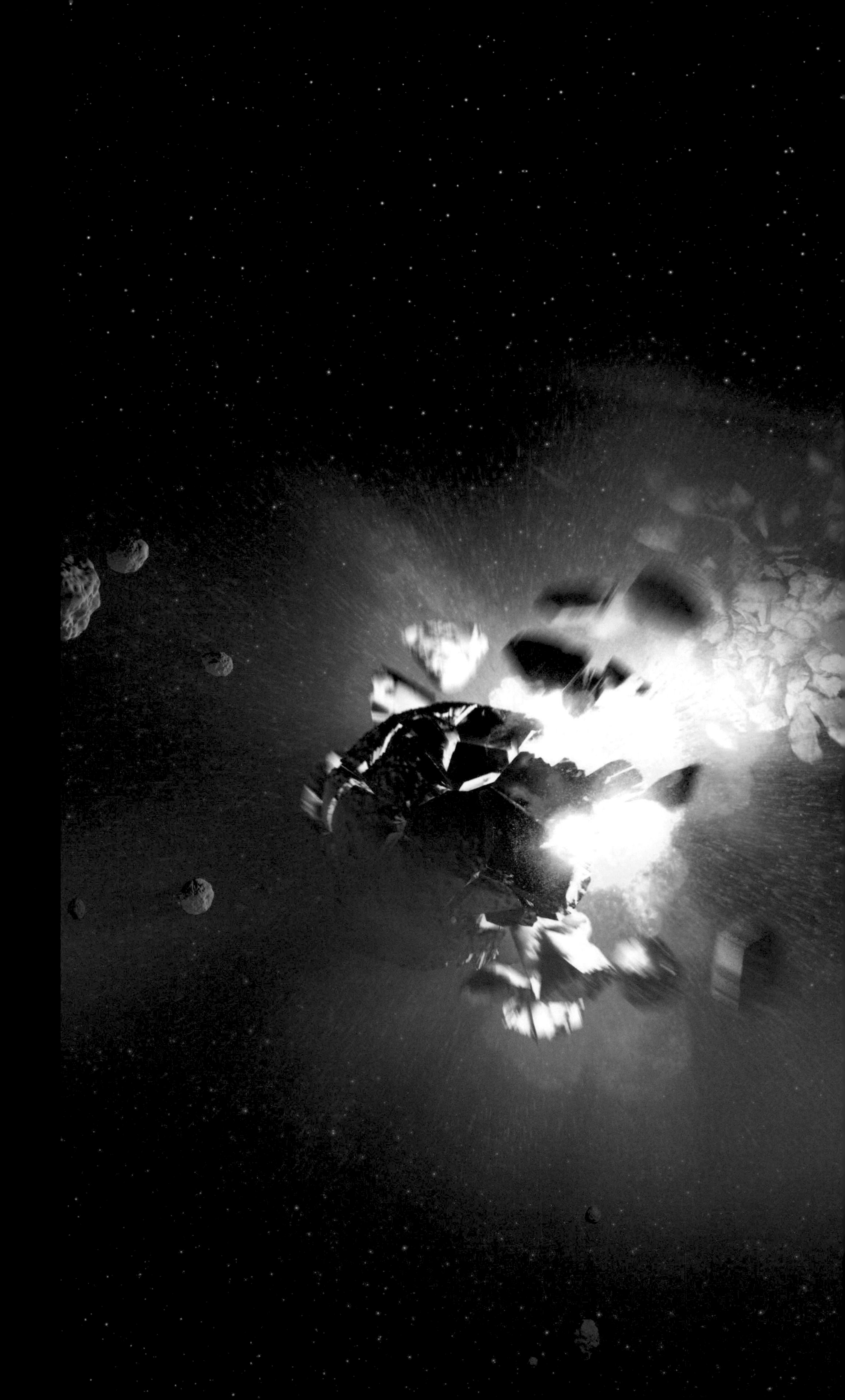

Right: Meteoroids are fragments and debris in space resulting from collisions among asteroids, comets, moons, and planets. This artist re-creation shows what a collision in deep space might look like.

[Picture and caption credit: NASA/JPL-Caltech]

Asteroids

Asteroids are relatively small rocks that scientists say are objects remaining from the creation of the solar system. Estimates are that more than 1,113,000 exist, especially in the Asteroid Belt in the regions between the planets Mars and Jupiter. The largest known asteroid is Vesta, at about 329 miles in diameter.

Below: This Hubble Space Telescope image reveals the gradual self-destruction of an asteroid whose ejected dusty material has formed two long, thin, comet-like tails. The longer tail stretches more than 500,000 miles (800,000 kilometers) and is roughly 3,000 miles (4,800 kilometers) wide. The shorter tail is about a quarter as long. The streamers will eventually disperse into space. These unusual, transient features are evidence that the asteroid, known as (6478) Gault, is beginning to come apart by gently puffing off material in two separate episodes. Hubble's sharp view reveals that the tails are narrow streamers, suggesting that the dust was released in short bursts, lasting anywhere from a few hours to a few days.

[Photo and caption credit: NASA]

Above: This image is from the last sequence of images of the giant asteroid Vesta that NASA's Dawn spacecraft obtained while looking down at Vesta's north pole as it was departing. When Dawn arrived in July 2011, Vesta's northern region was in darkness. After more than a year at Vesta, the sunlight has now made it to Vesta's north pole, which is in the middle of the image.

[Photo and caption credit: NASA]

7: Future Space Expeditions

Since its inception, NASA's goal has been to explore the unknown and positively influence the future. Everyday life has been shaped by the American space agency, as evidenced by hand-held computers, cell phones, and navigation using the Global Positioning System. Since 2000, astronauts and cosmonauts from many nations have worked side-by-side aboard the International Space Station. NASA is now on the verge of launching the largest rocket ever built to continue pioneering exploration of the Moon and beyond.

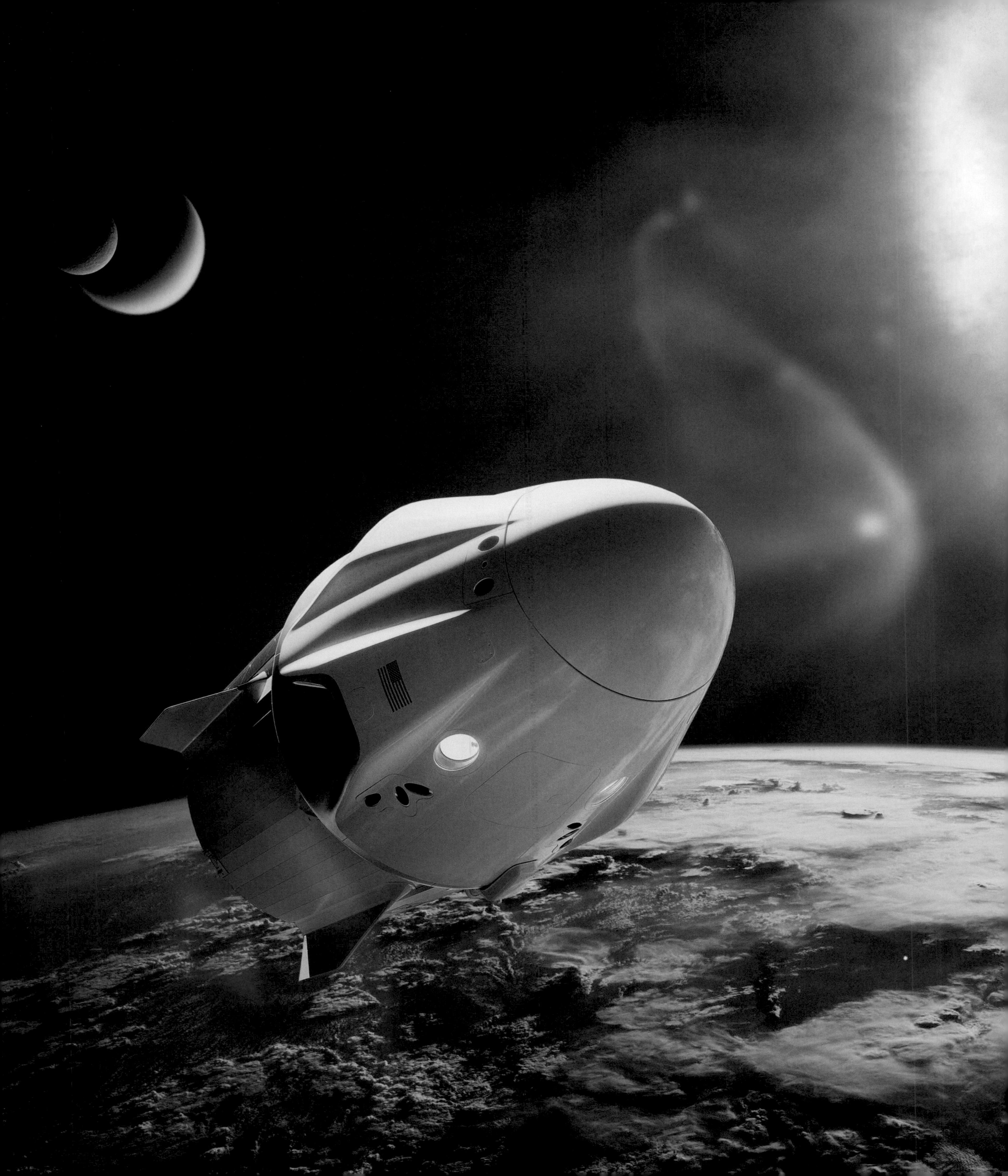

Human Spaceflight

Much can be accomplished with robotic probes traveling beyond Earth orbit. However, it is only when men and women venture into the unknown that uncharted territories can be truly understood. Only a pilot can make quick, last-second adjustments to a planned trajectory. When Apollo 11 landed in the Moon's Sea of Tranquility, Neil Armstrong took control, steering past a boulder-filled crater to land successfully.

For more than two decades, cutting-edge research and technology development has been taking place in the laboratories of the International Space Station. The unprecedented partnership includes the United States, Russia, Canada, Japan, and the nations of the European Space Agency.

NASA is now relying on industry to provide transportation to the orbiting outpost so the agency can focus on what it does best, exploration. Through the Artemis Program, NASA plans to establish a base on the Moon and, from what is learned on the lunar surface, send the first explorers to Mars.

Above: A pair of US spacesuits are pictured inside the International Space Station's Quest airlock, ready for an upcoming spacewalk.

[Photo and caption credit: NASA]

"One test result is worth one thousand expert opinions."

Wernher von Braun, German-American aerospace engineer

Above: This image from 24 April 2021, shows the SpaceX Crew Dragon Endeavour as it approached the International Space Station less than one day after launching from Kennedy Space Center in Florida. Crew Dragon is a privately owned spacecraft, developed and manufactured by SpaceX, an American aerospace manufacturer which was founded by Elon Musk in 2002.

[Photo and caption credit: NASA]

BEAM

NASA is investigating different approaches to expansion of the space station as well as spacecraft designs for missions beyond low-earth orbit. New habitation concepts are being studied and tested, including BEAM (Bigelow Expandable Activity Module), developed in a partnership between Bigelow Aerospace and NASA. It was launched on 8 April 2016 on a Commercial Resupply Services mission to the space station. The lightweight module was added to the station's Tranquility Node using the station's robotic arm. After a three-year engineering evaluation, NASA confirmed that BEAM would stay attached through 2028, noting that the module exceeded expectations. The module is now used for cargo storage on the space station.

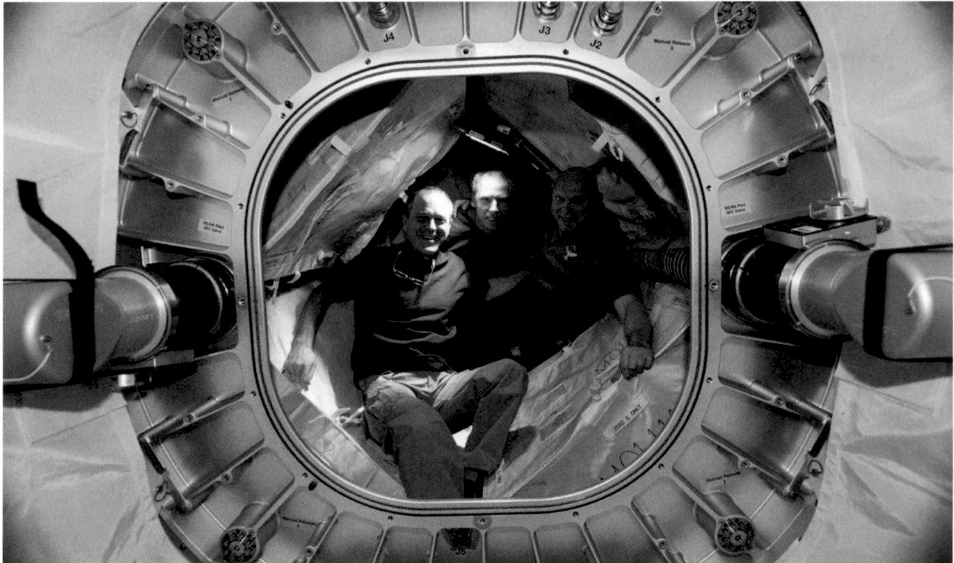

Top right: BEAM, the Bigelow Expandable Activity Module, was pictured installed to the Tranquility module with an external high definition camera.

Bottom right: Expedition 55/56 crew members (from left) NASA astronauts Ricky Arnold and Drew Feustel and Roscosmos cosmonaut Oleg Artemyev, pose for a portrait inside the Bigelow Expandable Aerospace Module (BEAM).

[Photo and caption credits: NASA]

Dream Chaser

Uncrewed resupply spacecraft provide a vital lifeline to the International Space Station, delivering needed supplies and equipment for on-going experiments. Cygnus spacecraft launched by Northrop Grumman Space Systems and the Dragon capsule, supplied by SpaceX, have been doing this for years. In the near future, Dream Chaser, a reusable spaceplane developed by Sierra Nevada Corporation, will also begin making deliveries. Plans call for Dream Chaser to lift off atop United Launch Alliance's Vulcan Centaur rocket, now being developed. Dream Chaser will rendezvous and dock with the space station, and will then return equipment to Earth, landing on a conventional runway.

Below: A full-scale engineering test article of Sierra Nevada Corporation Dream Chaser spacecraft is nearing completion for atmospheric evaluations at NASA Armstrong Flight Research Center at Edwards Air Force Base, CA.
[Photo and caption credit: NASA/Alamy Stock Photo]

Artemis Program

NASA's Artemis Program is designed to return astronauts to the Moon and, eventually, expand human presence to destinations farther into the solar system. The program gets its name from Artemis of Greek mythology, the twin sister of Apollo. The agency's Artemis Program aims to land the first woman and the first person of color on the lunar surface.

The program is a collaboration between NASA and its industry and international partners to establish a continuing presence on the Moon. Plans call for the Artemis Program to include assembly of Gateway, a small space station orbiting the Moon. Gateway is being designed to support operations on the lunar surface and provide a staging point for further exploration.

To return to the Moon, NASA has built the Space Launch System (SLS), the largest, most powerful rocket in history. At 322-feet-tall, the four engines of the SLS core stage, supplemented by twin solid rocket boosters, will generate 8.8 million pounds of thrust.

Above: Technicians are practicing by putting on the Self-Contained Atmospheric Protective Ensemble (SCAPE) suits for a test simulation of loading propellants into a replicated test tank for Orion. Exploration Ground Systems is preparing for Artemis I with a series of hazardous hyper test events at the MPPF.

[Photo and caption credit: NASA/Isaac Watson]

Left: Amy Ross, an advanced spacesuit designer at NASA's Johnson Space Center in Houston, stands with the Z-2, a prototype spacesuit. She has developed suits for the Moon and Mars. In 2020, the Perseverance Mars rover carried the first samples of spacesuit material ever sent to the Red Planet.
[Photo and caption credit: NASA/Alamy Stock Photo]

Space Launch System

The SLS rocket is designed to launch Orion, the first spacecraft developed to send humans beyond low-Earth orbit since the Apollo Program that ended in the 1970s. The new spacecraft will transport a crew of four on missions to the Moon. The first uncrewed flight test is planned for late 2022. An Orion will travel 280,000 miles from Earth, on a three-week mission, remaining in space longer than any vehicle designed for humans without docking to a space station.

Orion's next flight will transport astronauts on a lunar mission testing the spacecraft's systems for the first human flight to the Moon in more than 50 years. This will lead to returning crews to the lunar surface for extended periods of exploration, as well as future missions to destinations such as Mars.

Below: The core stage of the Space Launch System (SLS) rocket for NASA's Artemis I mission has been placed on the mobile launcher at Kennedy Space Center in Florida.
Opposite: Earth's Moon is seen rising behind NASA's Space Launch System rocket.
[Photo and caption credits: NASA]

NASA
NASA
esa

Parker Solar Probe

NASA's Parker Solar Probe is on a seven-year trip to complete the first mission for a close-up study of the Sun. The size of a compact car, the spacecraft traveled about 4 million miles to dive directly into the atmosphere of Earth's nearest star. Launched on 12 August 2018 on a Delta IV rocket, the Parker Solar Probe is making seven passes by the Sun, leading to the closest approach in 2025 while traveling at 430,000 mph, faster than any human-made object in history. Scientists believe they will be able to trace the energy in the Sun's corona, the outer layer of the star's atmosphere, and the solar wind. In doing so, researchers believe they will better understand the solar wind's magnetic fields that can affect life on Earth.

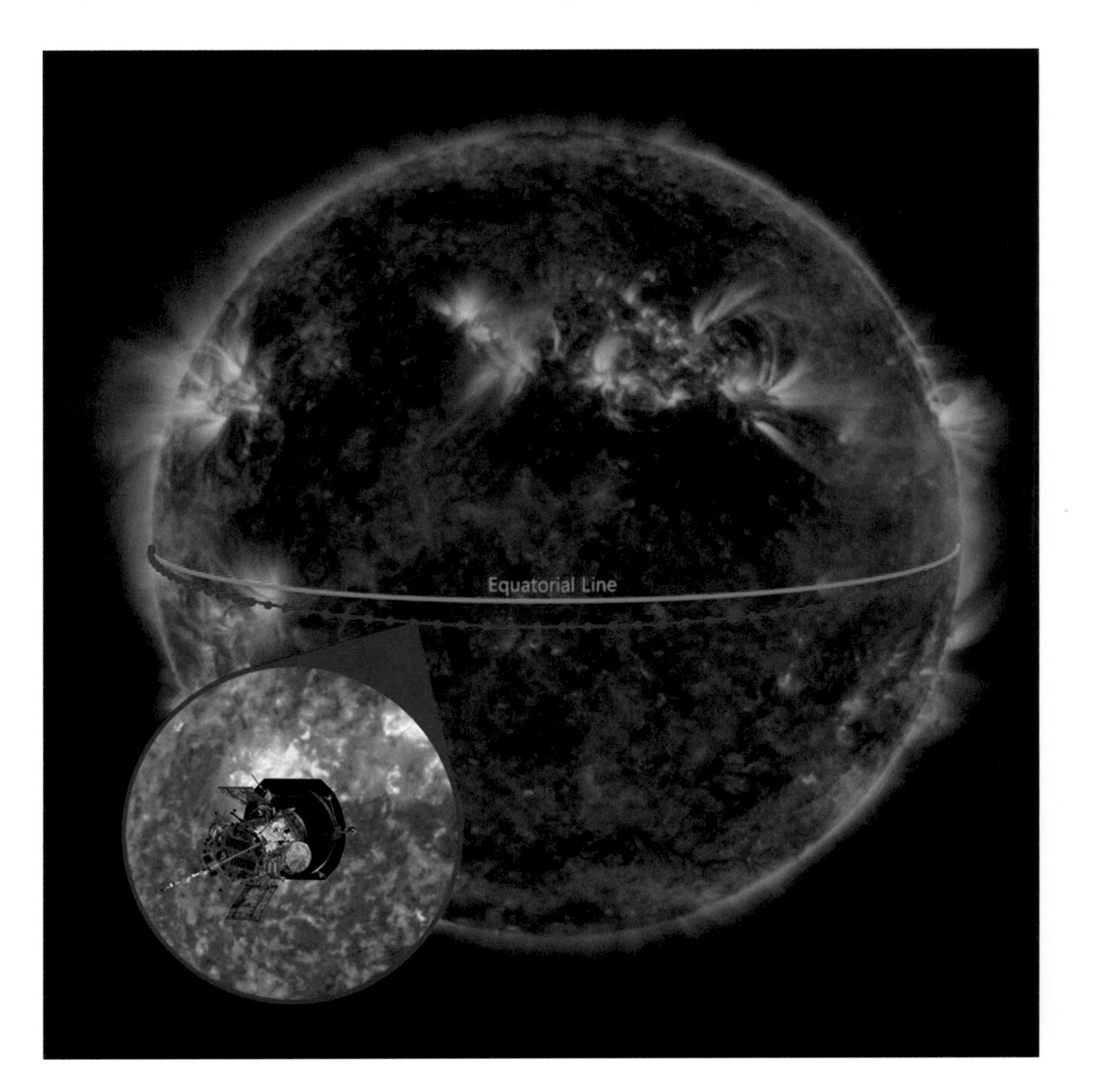

Right: The red line indicates the path of NASA's Parker Solar Probe across the face of the Sun, as seen from Earth on 24–27 February 2022. The red dots indicate an hour along the trajectory, and the appearance of the path heading into the Sun on the right accounts for Earth's own movement around our star. The image of the Sun was captured by NASA's Solar Dynamics Observatory.

[Photo and caption credit: NASA/Johns Hopkins APL/ Steve Gribben/SDO]

Europa Clipper

In October 1989, the NASA probe, Galileo, was launched from Space Shuttle Atlantis and spent eight years orbiting and studying the planet Jupiter. Plans call for Europa Clipper to lift off in 2024 on a SpaceX Falcon Heavy rocket. Europa Clipper is a follow-on to investigate Jupiter's moon, Europa. It will do this by using a series of fly-bys while orbiting this large planet.

Once Europa Clipper enters orbit around Jupiter, it will spend the next year adjusting its path to fly past Europa. The spacecraft will pass the large moon 45 times over a three-year period, transmitting imagery back to researchers on Earth. Scientists believe this will create a scan of most of Europa's surface that many believe could have conditions to sustain life beneath its icy crust. The mission is expected to also complement data returned from the European Space Agency's Jupiter Icy Moons Explorer set to launch in 2023, flying past Europa, Callisto, and Ganymede, all moons of Jupiter.

Below: This illustration, updated in December 2020, depicts NASA's Europa Clipper spacecraft. With an internal global ocean twice the size of Earth's oceans combined, Jupiter's moon Europa may have the potential to harbor life. The Europa Clipper orbiter will swoop around Jupiter on an elliptical path, dipping close to the moon on each fly-by to collect data.

[Photo and caption credit: NASA/JPL-Caltech]

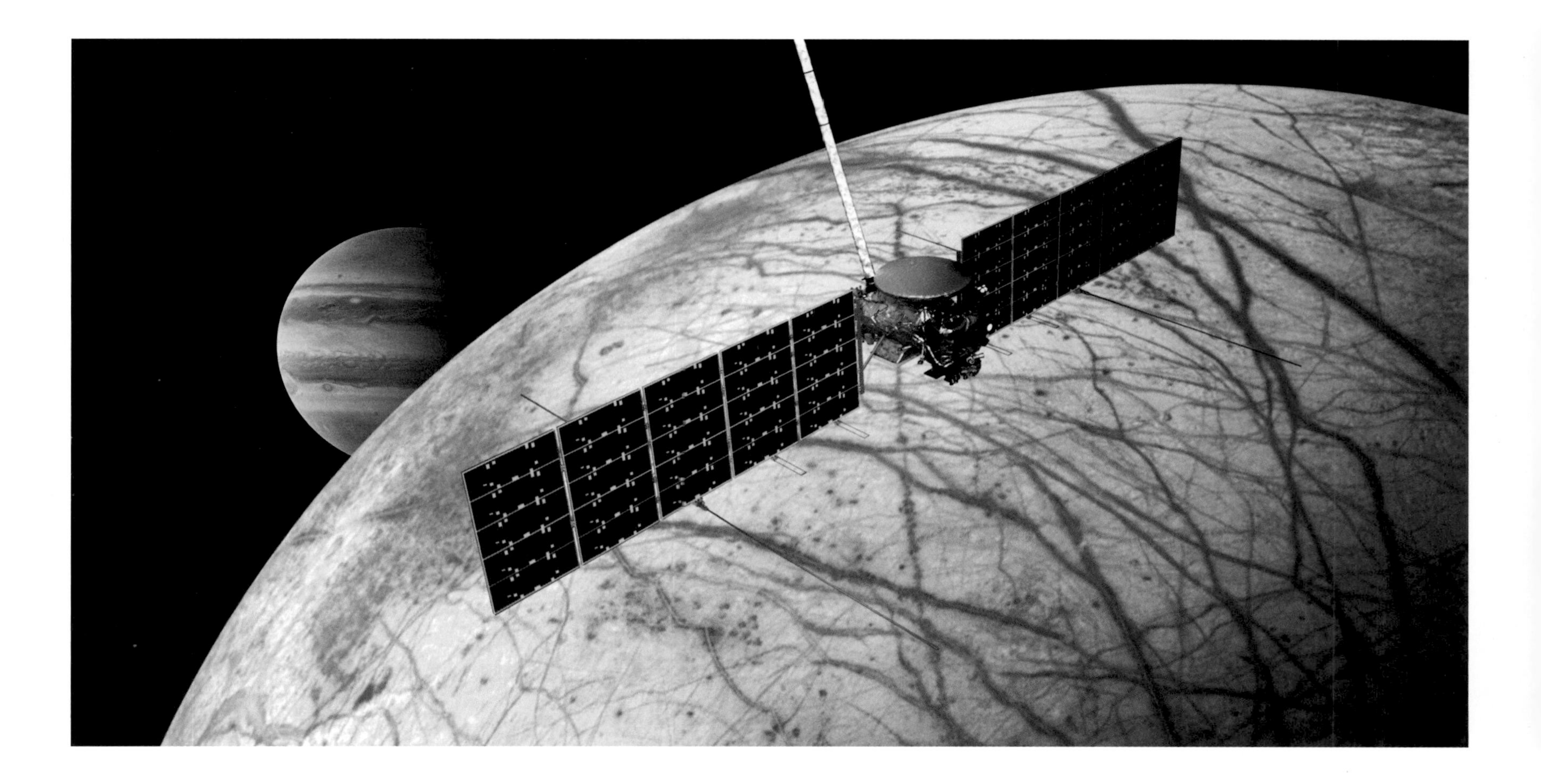

NASA's Great Observatories

For decades astronomers dreamed of placing a large telescope in space above the distortion of the Earth's atmosphere. In 1946, physicist Lyman Spitzer, proposed a space-based telescope in his scholarly paper, "Astronomical Advantages of an Extraterrestrial Observatory." The vision was realized when the crew of Space Shuttle Discovery deployed the Hubble Space Telescope on 25 April 1990. It was the first of NASA's Great Observatories.

Hubble Space Telescope

During its more than 30 years of continuing operations, the Hubble Telescope has helped astronomers re-write text books with its stunning images of planets, stars, galaxies, and other phenomena beyond Earth. Scientists note that while the telescope resolved many long-standing questions, its discoveries have raised new ones. Studies conducted by Hubble have helped determine the predominance of black holes in nearby galaxies and set the estimated age of the universe at about 13.7 billion years.

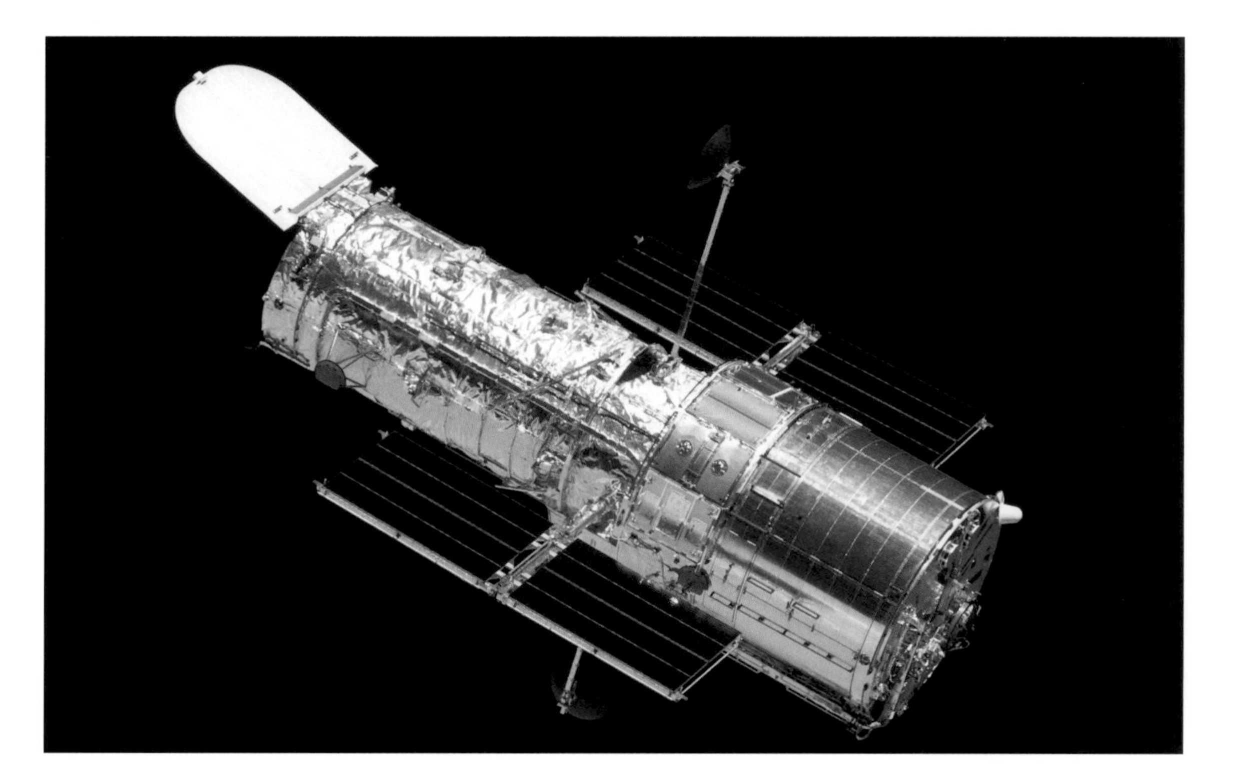

Right: Hubble Space Telescope seen from Space Shuttle Atlantis after servicing mission 4—the sixth and final Hubble mission in 2009.

[Photo and caption credit: NASA/Alamy]

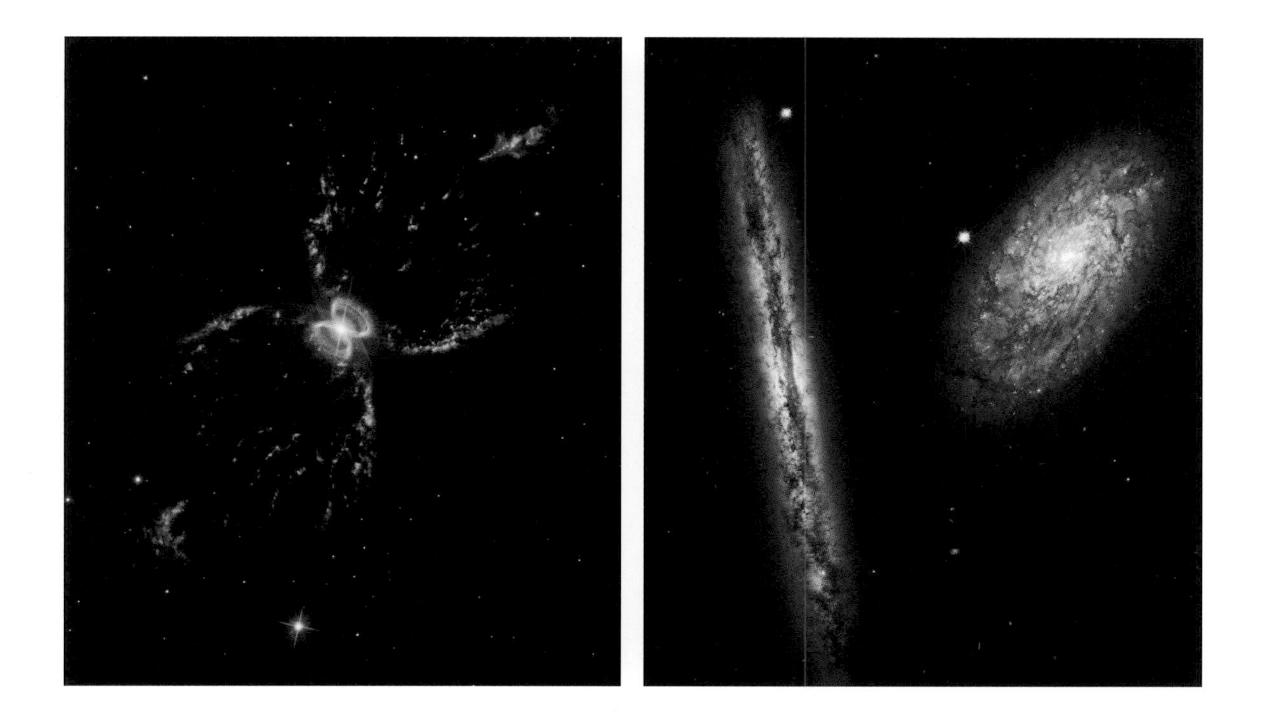

Far left: Southern Crab Nebula

Left: Galaxies NGC 4302 and NGC 2498 seen side on. Both are 55 million light years away from Earth.

Hubble's Deep Field, Ultra-Deep Field and Extreme Deep Field instruments have captured photographs of galaxies that are billions of light years from Earth. These images have depicted galaxies during the universe's earliest stages of formation.

In 2009, the Hubble Telescope acquired a colorful view of the Southern Crab Nebula. The large interstellar cloud of gas and dust has "hourglass-shaped" structures created by two rotating binary stars—one is an old red giant and the other a white dwarf.

As Hubble continues operating, scientists and astronomers photographed the comet Bernardinelli–Bernstein in April 2022. It was determined to be the largest comet nucleus ever observed with a mass estimated to be 50 times larger than any other known comet in the solar system.

"Earth is the cradle of humanity, but one cannot live in a cradle forever."

Konstantin Tsiolkovsky, Russian rocket scientist

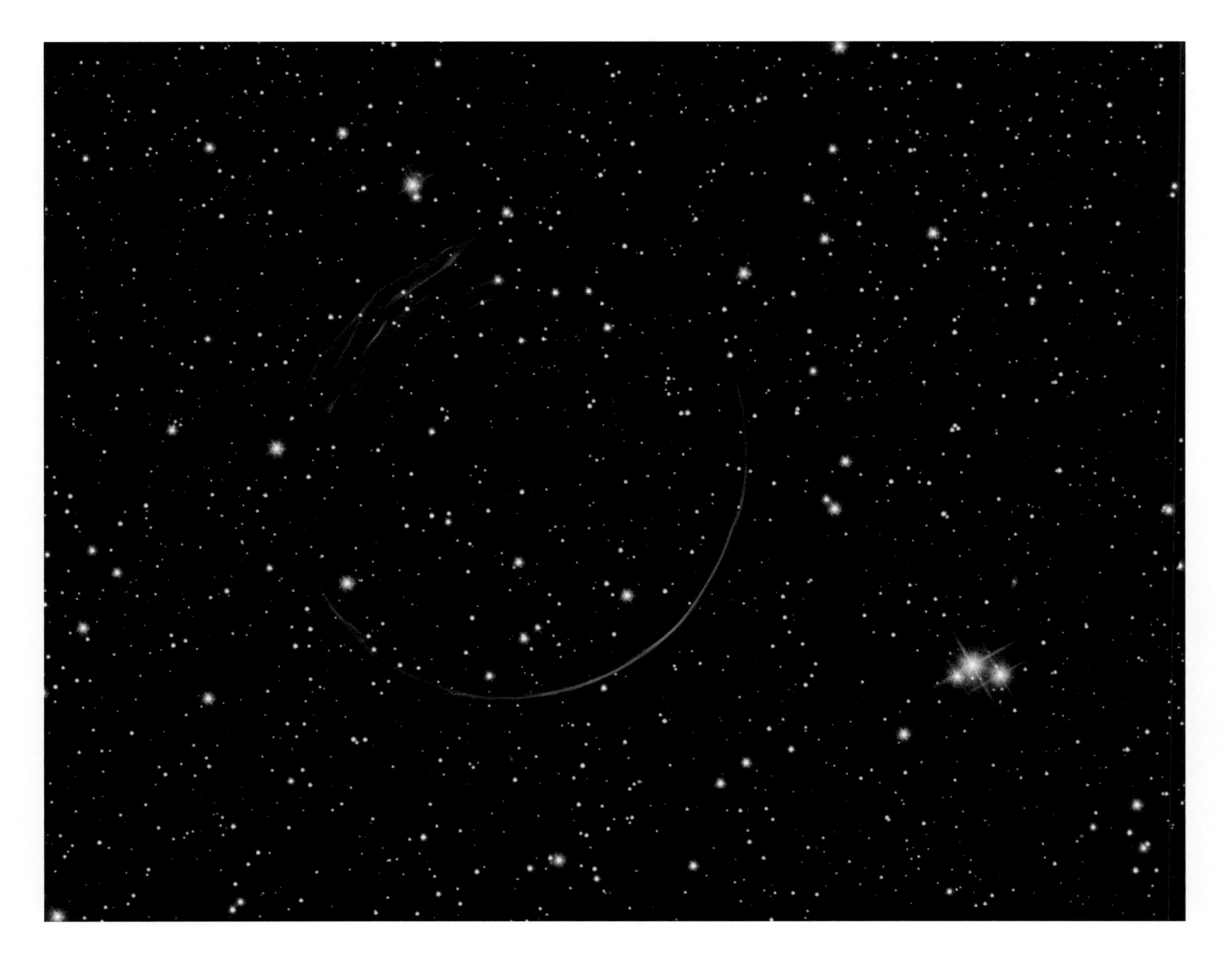

Above: This Supernova Bubble, captured by Hubble, is a ring of gas that is being shocked by the expanding blast wave from a supernova. Although it looks fairly small, it is actually 23 light-years across and expands at more than 11 million miles per hour.

[Photo credit: NASA]

Chandra X-ray Observatory

Launched in 1999 by the STS-93 crew of Space Shuttle Columbia, the Chandra X-ray Observatory provides astronomers with images of striking surroundings to better comprehend the evolution and make-up of the universe. Chandra's instruments are so sensitive that it can detect an object 100 times fainter than earlier instruments. Its mission is ongoing.

One of the objects studied by the Chandra Observatory is a 400-year-old supernova named for Tycho Brahe, the Danish astronomer who discovered it. A supernova is an exceptionally bright explosion of a star that creates a bright source of X-rays. Studies of the Tycho Supernova

Below: The Chandra X-Ray Observatory (Chandra XRO) studies black holes, quasars at the edge of the observable universe, and analyzes comets in our own solar system. Chandra is the most powerful X-ray telescope ever built, with eight times greater resolution and the ability to detect sources more than 20 times fainter than any previous X-ray telescope.
[Photo and caption credit: NASA]

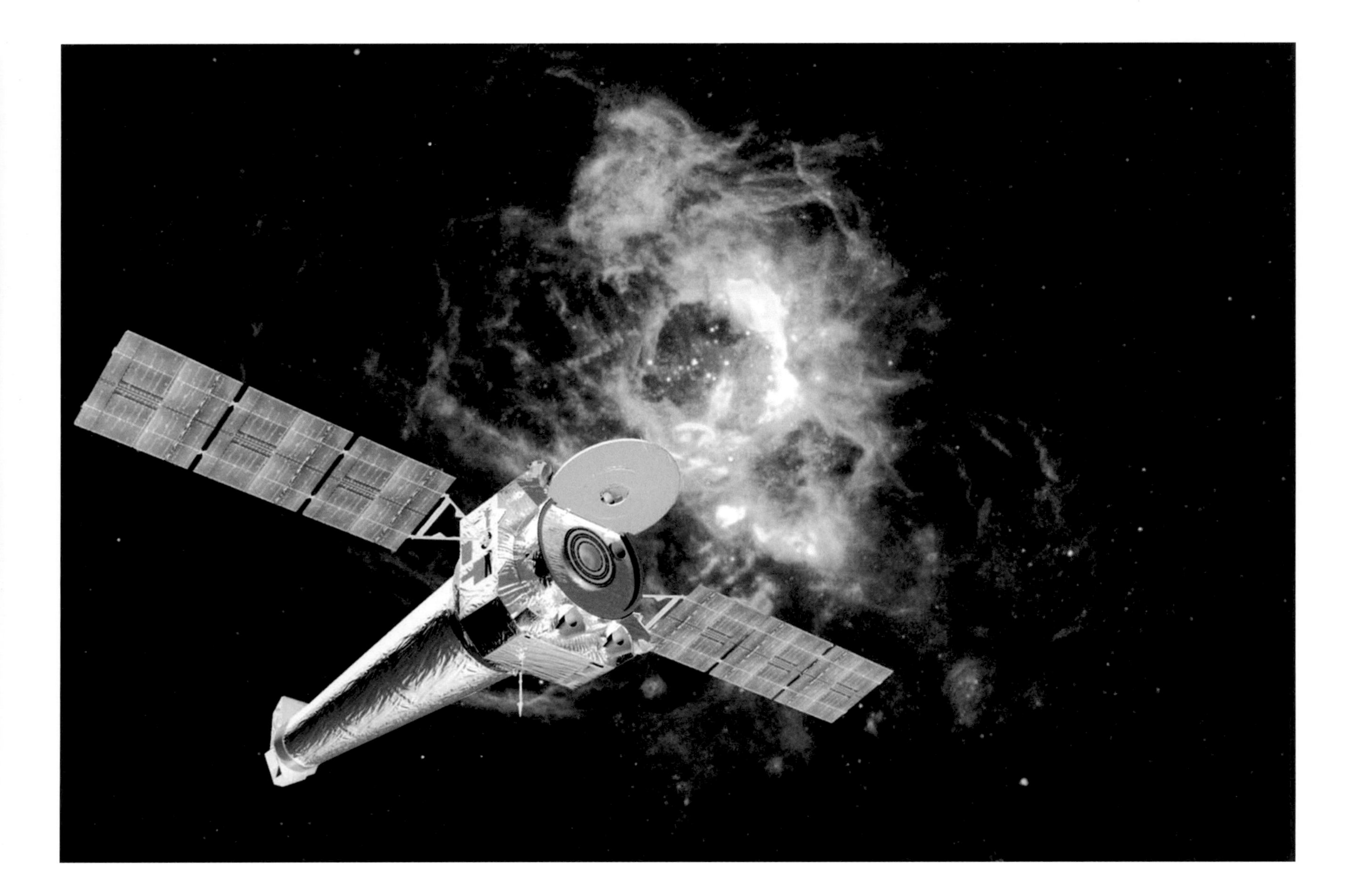

exposes the dynamics of the exploding star in astonishing detail. Information gathered by Chandra's instruments indicate that shock waves may have caused energetic particles, called cosmic rays, to saturate the Milky Way galaxy with effects on Earth.

NASA's Chandra X-ray Observatory is also proving especially useful in helping better understand the phenomena known as "black holes." When a star is dying, its gravity is compressed into a small space. The result is that the pull of gravity is so strong, even light cannot escape. It is believed that the gravity in black holes is pulling in thousands of stars, increasing its mass. Using Chandra, astronomers have completed research that has determined that there are two types of black holes. A lesser variety are "stellar-mass" black holes. These weigh 5 to 30 times that of the Sun. However, massive black holes exist near the center of many large galaxies and can weigh millions or even billions of times that of the Sun.

Below: Tycho Supernova remnant, first observed and studied by Danish astronomer Tycho Brahe in 1572.

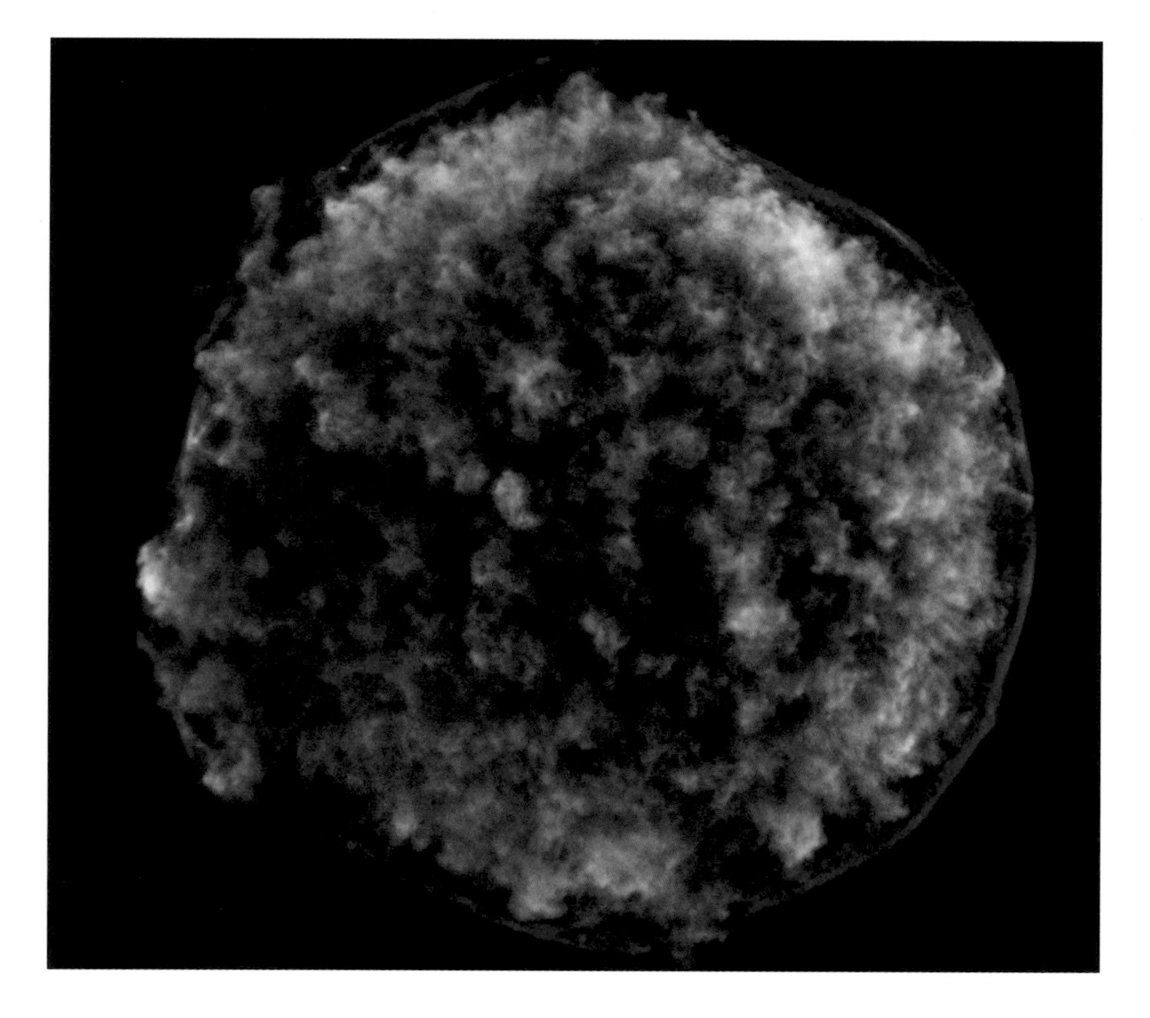

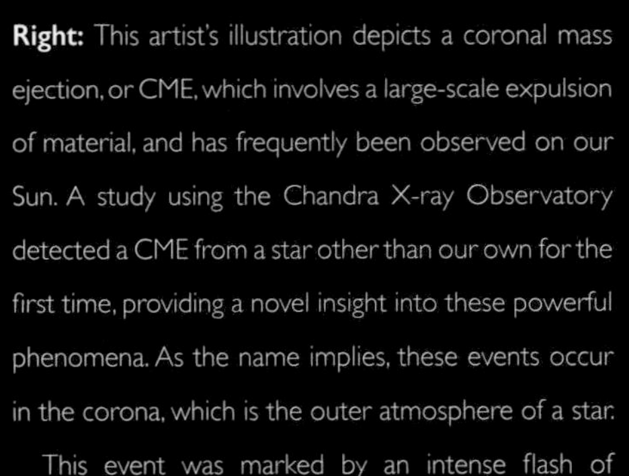

Right: This artist's illustration depicts a coronal mass ejection, or CME, which involves a large-scale expulsion of material, and has frequently been observed on our Sun. A study using the Chandra X-ray Observatory detected a CME from a star other than our own for the first time, providing a novel insight into these powerful phenomena. As the name implies, these events occur in the corona, which is the outer atmosphere of a star.

This event was marked by an intense flash of X-rays followed by the emission of a giant bubble of plasma (hot gas containing charged particles).

[Photo and caption credit: NASA/CXC/INAF/ Argiroffi, C. et al./S. Wiessinger]

Spitzer Space Telescope

The Spitzer Space Telescope was launched in 2003 by a Delta II rocket into an "Earth-trailing" orbit. While the Hubble Space Telescope circles the Earth, NASA's Spitzer orbits the Sun, following Earth in its path but at a slower pace. The advantage is that Spitzer was less affected by Earth's heat and had a wider field to observe objects in space.

The Spitzer Telescope's infrared instruments are designed to view objects detecting heat from targets in space. This allowed astronomers to study phenomena such as newly forming planetary systems not visible to optical telescopes. Spitzer also focused on "stellar nurseries," areas in a dense nebula in which new stars are forming. Due to loss of the spacecraft's coolant in 2009, the Spitzer Space Telescope was shut down and set to a safe-mode in January 2020.

Below: An artist's impression of the Spitzer Space Telescope. It was designed to detect infrared radiation, which is primarily heat radiation, allowing scientists to examine cosmic regions that are hidden from optical telescopes, such as dusty stellar nurseries, the centers of galaxies, and newly formed planetary systems.
[Picture and caption: NASA]

TRAPPIST-1

Using the Spitzer Space Telescope, astronomers discovered an exoplanet, similar in size to the Earth. It exists in a "habitable zone"—a region close enough to a host star that the temperature could support liquid water. It orbits a star that includes seven "rocky" planets, similar in makeup to Earth and Mars. As scientists scrutinize similar exoplanets, they may find planets with the potential for sustaining life as we know it.

Below: This illustration shows what the TRAPPIST-1 system might look like from a vantage point near planet TRAPPIST-1f (at right)
[Picture and caption credit: NASA/JPL-Caltech]

"Space exploration is vital. If we fail to save the Earth we will need a backup planet."

Dave Bishop, friend of the author

Above: This artist's concept shows a brown dwarf surrounded by a swirling disk of planet-building dust. NASA's Spitzer Space Telescope spotted such a disk around a surprisingly low-mass brown dwarf, or "failed star." Astronomers believe that this unusual system will eventually spawn planets. If so, they speculate the disk has enough mass to make one small gas giant and a few Earth-sized rocky planets. [Picture and caption credit: NASA/JPL]

James Webb Space Telescope

Named for NASA's second administrator, the James Webb Space Telescope is a NASA-led collaboration with space agencies in Europe and Canada. The Webb telescope was launched on Christmas Day in 2021 on an Ariane 5 rocket lifting off from Guiana Space Centre in Kourou, French Guiana. It is now operational in an orbit circling around a point 930,000 miles beyond Earth.

The most powerful telescope ever launched, the Webb telescope is expected to open new chapters in astronomy. It has enhanced infrared capabilities, allowing its instruments to observe stars, galaxies, and other phenomena too far way or faint for Hubble. Because the Webb telescope will be able to see such distant objects, it should be able to probe the makeup of the universe by seeing galaxies as they were during their formation.

Above: The James Webb Space Telescope is the scientific successor to NASA's Hubble Space Telescope. In this photo, NASA technicians lifted the telescope using a crane and moved it inside a clean room at NASA's Goddard Space Flight Center in Greenbelt, Maryland. Once launched into space, the Webb telescope's 18-segmented gold mirror is specially designed to capture infrared light from the first galaxies that formed in the early universe and will help the telescope peer inside dust clouds where stars and planetary systems are forming today.

[Photo and caption credit: NASA]

Exoplanets

Not long ago, astronomers could only wonder if planets similar to Earth existed beyond the solar system. But new worlds have now been pinpointed for further study through NASA's Kepler Space Telescope, launched on 7 March 2009, and the agency's Transiting Exoplanet Survey Satellite (TESS), which lifted off on 18 April 2018. Scientists have now discovered that many stars may have one or more exoplanet, a planet located outside the solar system.

Scientists have also established that exoplanets can have a rocky surface similar to Earth or Mars. However, they also can be gas giants similar to Jupiter. Some circle a star so closely that their orbit only lasts a few days, creating a short "year." On the other hand, astronomers have determined that some exoplanets do not circle any particular star. Instead, they simply meander through their galaxy.

Kepler-186f

An exoplanet designated "Kepler-186f" was the first found by the Kepler Space Telescope to be similar in size to Earth. It is significant because astronomers say that it is in a "habitable zone"—a region close enough to a host star that the temperature could support liquid

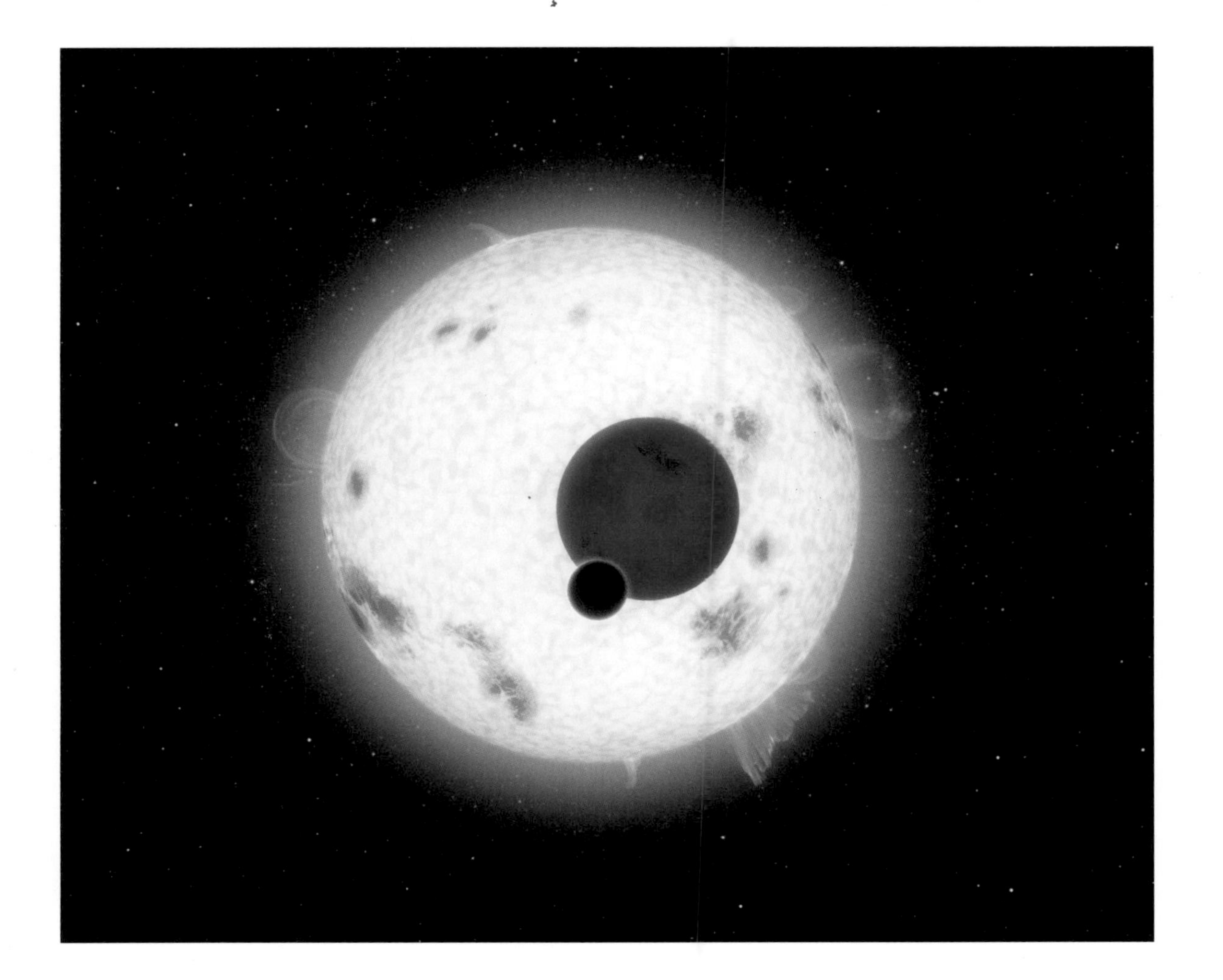

Left: This artist's concept shows Kepler-16b with its two suns.

[Picture credit: NASA/JPL-Caltech/R. Hurt]

Opposite: This artist's concept depicts Kepler-186f, the first validated Earth-size planet to orbit a distant star in the habitable zone.

[Picture and caption credit: NASA Ames/SETI Institute/JPL-Caltech]

water, a key element to support life as we know it. While current technology does not allow scientists to see if anyone is looking back at Earth, for the first time astronomers are able to confirm that there are other worlds in the universe beyond the solar system.

Kepler-16b

The Kepler telescope also found a planet that orbits a binary star system meaning that it has two suns. Fans of *Star Wars* will recognize that Luke Skywalker's fictitious home planet of Tatooine had two suns. However, evidence points to Kepler-16b being an inhabitable, cold world. Both of the stars are much smaller than Earth's Sun. The larger of the two is only about 69 percent the size of the star at the center of the solar system and the smaller is 20 percent that of the Sun.

Thank you: I give special thanks to Yvette Smith, from NASA, for being so professional, profoundly patient and straightforward in helping me through the whole process of putting this book together. For Carolyn Robb whose research, writing skills and perfectionism served well for this book. To Robert Granath for his invaluable historic expertise about NASA. With much gratitude and appreciation to the key executives at ACC Art Books; Alice Bowden; Senior Editor, whose ability to edit is only matched by her ability to have thought-provoking debates resulting in a book that exceeds my expectation, to Andrew Whittaker; Senior Editor; for converting the 'big picture' into reality, and most importantly to the CEO; James Smith; whose taste in authors is impeccable.

Picture Credits:
Front and back cover: NASA
Pages 2, 10, 11, 60, 61, 101, 156, 157, 186, 187: NASA.
Pages 231, 269: Alamy.
All other pages credited individually.

ISBN: 978-178884-179-5

British Library Cataloguing-in-Publication Data
A catalogue record for this book is available from the British Library

Expert Consultant: Robert Granath
Senior Editor: Alice Bowden
Production: Craig Holden

Printed in Belgium
for ACC Art Books Ltd., 6 West 18th Street, 4B, New York, NY 10011

www.accartbooks.com

ACC
ART
BOOKS